MAKING SENSE OF MATHEMATICS FOR TEACHING

to Inform Instructional Quality

MELISSA D. BOSTON
AMBER G. CANDELA
JULI K. DIXON

Solution Tree | Press

Copyright © 2019 by Solution Tree Press

Materials appearing here are copyrighted. With one exception, all rights are reserved. Readers may reproduce only those pages marked "Reproducible." Otherwise, no part of this book may be reproduced or transmitted in any form or by any means (electronic, photocopying, recording, or otherwise) without prior written permission of the publisher.

555 North Morton Street
Bloomington, IN 47404
800.733.6786 (toll free) / 812.336.7700
FAX: 812.336.7790

email: info@SolutionTree.com
SolutionTree.com

Visit **go.SolutionTree.com/mathematics** to download the free reproducibles in this book.

Printed in the United States of America

Library of Congress Cataloging-in-Publication Data

Names: Boston, Melissa, author. | Candela, Amber G., 1981- author. | Dixon, Juli K., author.
Title: Making sense of mathematics for teaching to inform instructional quality / authors: Melissa D. Boston, Amber G. Candela, and Juli K. Dixon.
Description: Bloomington, IN : Solution Tree Press, [2019] | Includes bibliographical references and index.
Identifiers: LCCN 2018040092 | ISBN 9781947604094 (perfect bound)
Subjects: LCSH: Mathematics--Study and teaching.
Classification: LCC QA11.2 .B673 2019 | DDC 510.71--dc23 LC record available at https://lccn.loc.gov/2018040092

Solution Tree
Jeffrey C. Jones, CEO
Edmund M. Ackerman, President

Solution Tree Press
President and Publisher: Douglas M. Rife
Associate Publisher: Sarah Payne-Mills
Art Director: Rian Anderson
Managing Production Editor: Kendra Slayton
Production Editor: Alissa Voss
Senior Editor: Amy Rubenstein
Copy Editor: Miranda Addonizio
Proofreader: Jessi Finn
Text and Cover Designer: Abigail Bowen
Editorial Assistant: Sarah Ludwig

"Elapsed Time Lesson" and "Decimals on a Number Line" videos licensed by University of Central Florida Research Foundation, Inc. Copyright © 2017 University of Central Florida Research Foundation, Inc. All Rights Reserved.

Acknowledgments

I want to thank my husband, David, and daughters, Lauren and Lindsey, who inspire and inform the quality of everything I do, and to express my sincere appreciation to everyone who has participated in the journey of the IQA throughout the years.

—Melissa Boston

I would like to express gratitude for my family and friends who have supported me every step of the way. I would also like to thank each and every student, teacher, and colleague I have ever taught or interacted with; you have inspired me daily and informed the quality of my instruction.

—Amber Grace Candela

My deepest love and gratitude to my family for informing my instructional quality on a daily basis and for always supporting my endeavors, wherever they might take us: my daughters, Alex and Jessica; my husband, Marc; and my parents, Joy and Harvey.

—Juli Dixon

Thank you to the team at DNA Mathematics for supporting us to build on the work of the *Making Sense of Mathematics for Teaching* series. Thanks also to Alissa Voss for ensuring that our message is clear, and to Sarah Payne-Mills for continuing that support. A special thanks to the students in each of the videos we shared. Much appreciation to Jeff Jones and Douglas Rife for supporting and investing in our vision. To our reviewers, thank you for taking the time to think deeply and provide such helpful feedback about our work.

—Melissa Boston, Amber Grace Candela, and Juli Dixon

Solution Tree Press would like to thank the following reviewers:

Kristopher J. Childs
Assistant Professor of STEM Education
Texas Tech University
Lubbock, Texas

Samuel Martinez
Elementary Math Manager II
Fresno Unified School District
Fresno, California

Nicole Rigelman
Professor of Mathematics Education
Portland State University
Portland, Oregon

Mike Schrimpf
Assistant Head of School
Premier Charter School
St. Louis, Missouri

AnnMarie Varlotta
Math Instructional Support Teacher of
 Secondary Mathematics
Howard County Schools
Howard County, Maryland

Mary Velez
Mathematics Teacher
Roy C. Ketcham High School
Wappingers, New York

Visit **go.SolutionTree.com/mathematics** to download the free reproducibles in this book.

Table of Contents

Reproducible pages are in italics.

About the Authors ix

Introduction 1
    The Importance of Making Sense of Mathematics for Teaching 1
    About This Book 1

PART 1: Connecting to the *T* in TQE: Tasks and Task Implementation 5

CHAPTER 1
Potential of the Task 7
    Introductory Activities 7
        Activity 1.1: Solving a Task 7
        Activity 1.2: Considering Different Types of Tasks 10
    The IQA Potential of the Task Rubric 14
    Application Activities 16
        Activity 1.3: Rating Mathematical Tasks Using the Potential of the Task Rubric 16
        Activity 1.4: Using the IQA Potential of the Task Rubric to Rate and Adapt Tasks 21
    Considerations When Rating Tasks 24
        Defining the Task 25
        Considering Implications of Higher-Level Thinking 25
        Aligning With Learning Goals 26
    Summary 27

CHAPTER 2
Implementation of the Task 29
    Introductory Activities 29
        Activity 2.1: Comparing Two Mathematics Lessons 29
    The IQA Implementation of the Task Rubric 32
    Application Activities 34
        Activity 2.2: Using the Implementation of the Task Rubric 34

Activity 2.3: Identifying Evidence of Students' Thinking and Reasoning in Students' Work. . . . 38
Activity 2.4: Revisiting the Chapter 1 Transition Activity—Moving From Tasks
to Implementation. 45
Considerations for Implementation Based on Classroom Observations or Videos 46
Activity 2.5: Rating Implementation of the Task—Father and Son Race Lesson Version 1. 48
Activity 2.6: Rating Implementation of the Task—Father and Son Race Lesson Version 2 50
Summary. 52

PART 2: Connecting to the *Q* in TQE: Questions and Their Role as Discourse Actions 55

CHAPTER 3
Teacher's Questions 57

Introductory Activities 57
Activity 3.1: Identifying Different Types of Questions 57
Activity 3.2: Sorting Questions 61
Activity 3.3: Creating Questions 63
Activity 3.4: Revisiting the Chapter 2 Transition Activity—How Teacher's Questions Impact
Implementation 65
The IQA Teacher's Questions Rubric 67
Application Activities 68
Activity 3.5: Rating the Teacher's Questions in the 26 Divided by 4 Lesson 68
Summary. 70

CHAPTER 4
Teacher's Linking and Teacher's Press 73

Introductory Activities 73
Activity 4.1: Following Up on Students' Contributions—Teacher's Linking 73
Activity 4.2: Following Up on Students' Contributions—Teacher's Press. 76
Activity 4.3: Revisiting the Chapter 3 Transition Activity—Teacher's Questions and Follow-Up. . 79
The IQA Teacher's Linking and Teacher's Press Rubrics. 80
Application Activities 83
Activity 4.4: Rating Teacher's Linking and Teacher's Press 83
Activity 4.5: Determining When It Is Appropriate to Ask a Follow-Up Question 87
Summary. 89

PART 3: Connecting to the *E* in TQE: Evidence of Students' Mathematical Work and Thinking 91

CHAPTER 5
Students' Linking and Students' Providing 93

Introductory Activities 93
    Activity 5.1: Examining Students' Contributions—Students' Linking 93
    Activity 5.2: Examining Students' Contributions—Students' Providing 95
    Activity 5.3: Revisiting the Chapter 4 Transition Activity—Teacher's Linking, Teacher's Press, and Students' Contributions 97

The IQA Students' Linking and Students' Providing Rubrics 100

Application Activities 101
    Activity 5.4: Rating Students' Linking and Students' Providing 101

Summary 104

CHAPTER 6
The IQA Toolkit as a Tool to Assess and Improve Instructional Practice 107

Introductory Activities 107
    Activity 6.1: Rating a Small-Group Lesson 107
    Activity 6.2: Rating a Whole-Class Lesson 112

The IQA Rubrics as Tools to Reflect on and Improve Instruction 118

Application Activities 119
    Activity 6.3: Considering the Teacher's Reflection and Next Steps 119
    Activity 6.4: Revisiting the Chapter 5 Transition Activity—Using All IQA Rubrics 121

Summary 123

EPILOGUE
Next Steps 125

APPENDIX A
The IQA Toolkit 127

IQA Potential of the Task Rubric 128
IQA Implementation of the Task Rubric 129
IQA Teacher's Questions Rubric 130
IQA Teacher's Linking Rubric 131
IQA Teacher's Press Rubric 132
IQA Students' Linking Rubric 133

IQA Students' Providing Rubric *134*
IQA Implementation Observation Tool *135*
Framework for Different Types of Questions *136*

APPENDIX B
Suggested Answers for Activity 1.4 137

APPENDIX C
Suggested Answers for Activity 3.2 139

APPENDIX D
Additional Support for Rating Tasks 141
Level 1 141
Level 2 141
Level 3 142
Level 4 142

APPENDIX E
List of Figures and Videos 143

References and Resources 147

Index 151

About the Authors

Melissa D. Boston, EdD, is a professor of mathematics education at Duquesne University. She teaches mathematics methods courses for elementary, middle, and high school teachers, and previously taught similar courses at Slippery Rock University and the University of Pittsburgh. Earlier, she taught middle school and high school mathematics. Dr. Boston provides professional development and conducts research across the United States. She has obtained and directed grants for professional development projects from the Heinz Foundation, National Science Foundation, and Spencer Foundation, and she has served as external evaluator for three Mathematics–Science Partnership grants. In her research, Dr. Boston examines mathematics teaching and learning through classroom observations and collections of students' work using the Instructional Quality Assessment (IQA) in Mathematics Toolkit, for which she is lead developer.

Dr. Boston is a member of the National Council of Teachers of Mathematics (NCTM), Association of Mathematics Teacher Educators (AMTE), and Pennsylvania Association of Mathematics Teacher Educators (PAMTE). She has served on NCTM's Student Explorations in Mathematics committee, as associate editor of *Mathematics Teacher Educator* (2012–2015), as series editor for NCTM's 2017–2018 *Annual Perspectives in Mathematics Education*, on NCTM's *Principles to Actions Toolkit* development team, and on research advisory boards. She was also awarded the Association of Teacher Educators' 2008 Distinguished Dissertation Award.

Dr. Boston is lead author of NCTM's *Taking Action: Implementing Effective Mathematics Teaching Practices in Grades 9–12*. She has published her work in scholarly journals (*Elementary School Journal*, *Journal of Mathematics Teacher Education*, *Journal for Research in Mathematics Education*, *International Journal of Mathematics Teacher Education*, *Journal of Mathematics Education Leadership*, and *Urban Education*) and book chapters. Dr. Boston received a bachelor of science degree in mathematics and secondary mathematics education from Grove City College and a master of arts degree in mathematics and a doctorate in mathematics education from the University of Pittsburgh. To learn more about Dr. Boston's work, follow @MBostonMath on Twitter.

Amber G. Candela, PhD, is an assistant professor of mathematics education at the University of Missouri–St. Louis (UMSL). She currently teaches mathematics methods classes for prospective elementary, middle, and high school teachers in the teacher education program at UMSL. Previously, she taught middle school mathematics in both suburban and urban settings in Charlotte, North Carolina, and East Harlem, New York. Dr. Candela is focused on supporting teachers' implementation of cognitively challenging tasks so each and every student has access to high-quality mathematics. In 2016 she was honored with the Gitner

Excellence in Teaching Award at UMSL, and in 2018 she was honored with the Missouri Council of Teachers of Mathematics (MCTM) Outstanding Post-Secondary Educator Award.

Dr. Candela is an active member of the National Council of Teachers of Mathematics (NCTM) and TODOS Mathematics for All organizations. She has worked in schools providing professional development on selecting and implementing tasks and is dedicated to supporting each and every learner in the classroom and focusing on how tasks are implemented in inclusive settings.

Dr. Candela received a bachelor's degree in mathematics and education from St. Bonaventure University, a master's degree in mathematics education from Appalachian State University, and a doctorate in mathematics education from the University of Georgia. To learn more about Dr. Candela's work, follow @AmCan36 on Twitter.

Juli K. Dixon, PhD, is a professor of mathematics education at the University of Central Florida (UCF) in Orlando. Prior to joining the faculty at UCF, Dr. Dixon was a secondary mathematics educator at the University of Nevada, Las Vegas and a public school mathematics teacher in urban school settings at the elementary, middle, and secondary levels. Dr. Dixon is focused on improving teachers' mathematics knowledge for teaching so that they support their students to communicate and justify mathematical ideas.

She is a prolific writer who has authored and coauthored books, textbooks, chapters, and articles. She is also a lead author on the *Making Sense of Mathematics for Teaching* professional development book and video series as well as for the K–12 school mathematics textbooks *GO Math!*, *AGA*, *Integrated Math*, and *Into Math* with Houghton Mifflin Harcourt. Especially important to Dr. Dixon is the need to teach each and every student. She often shares her personal story of supporting her own children with special needs to learn mathematics in an inclusive setting. Dr. Dixon published *A Stroke of Luck: A Girl's Second Chance at Life* with her daughter, Jessica Dixon. A sought-after speaker, Dr. Dixon has delivered keynotes and other presentations throughout North America.

Dr. Dixon received a bachelor's degree in mathematics and education from the State University of New York at Potsdam, a master's degree in mathematics education from Syracuse University, and a doctorate in curriculum and instruction with an emphasis in mathematics education from the University of Florida. Dr. Dixon is a leader in DNA Mathematics.

To learn more about Dr. Dixon's work supporting children with special needs, visit www.astrokeofluck.net, or to learn more about Dr. Dixon's work supporting teachers, follow @thestrokeofluck on Twitter.

To book Melissa D. Boston, Amber G. Candela, or Juli K. Dixon for professional development, contact pd@SolutionTree.com.

Introduction

Effective teaching practice can be learned.
—National Research Council

Improving instructional quality is an important aspect of teaching mathematics effectively, as instructional quality in mathematics has been associated with student achievement (Darling-Hammond, 2000). Our vision of quality mathematics instruction aligns with the effective mathematics teaching practices described by the National Council of Teachers of Mathematics (NCTM, 2014) in *Principles to Actions: Ensuring Mathematical Success for All*. According to NCTM (2014), "Effective teaching of mathematics engages students in solving and discussing tasks that promote mathematical reasoning and problem solving and allow multiple entry points and varied solution strategies" (p. 17). In this book, we provide you with a process and toolkit for improving instructional quality that are specific to mathematics teaching and learning, that identify aspects of instruction that impact students' learning, and that provide data to influence the planning and teaching of future lessons.

The Importance of Making Sense of Mathematics for Teaching

Previous books in the *Making Sense of Mathematics for Teaching* series have focused on developing a deep understanding of important mathematical content at different grade bands. They provide opportunities for mathematics teachers in grades K–2, 3–5, 6–8, and high school to engage deeply with the mathematics they teach. A central premise in each of these books is that you learn about the mathematics you teach by doing the mathematics you teach (Nolan, Dixon, Roy, & Andreasen, 2016).

In this book, we focus on the opportunities for thinking and reasoning embedded in mathematical tasks at all grade levels and present an instructional assessment framework to determine whether those opportunities are actualized during instruction. Just as the previous books in the *Making Sense of Mathematics for Teaching* series promote mathematical learning by asking readers to *do* mathematics, this book engages teachers in reflecting on mathematics instruction through the activities within each chapter. Throughout the book, we ask readers to solve mathematical tasks. In each of those instances, readers should take time to solve the tasks and discuss their solution strategies within their collaborative teams. This will provide greater insight when they later analyze the task or the lesson featuring the task.

About This Book

Making Sense of Mathematics for Teaching to Inform Instructional Quality provides teachers and teacher leaders with the opportunity to reflect on the quality of mathematics instruction at any grade level. As with previous books in the *Making Sense of Mathematics for Teaching* series, this book highlights the use of tasks, questions, and evidence (referred to as the TQE process; see figure I.1). Part 1 engages teachers in considering the quality of *tasks* and task implementation (connecting to the *T* in the TQE process). Part 2 supports teachers as they explore the quality and impact of their *questions* and other discourse actions (the *Q* in the TQE process). Part 3 guides educators to examine the *evidence* of their students' thinking and participation in the classroom community (the *E* in the TQE process).

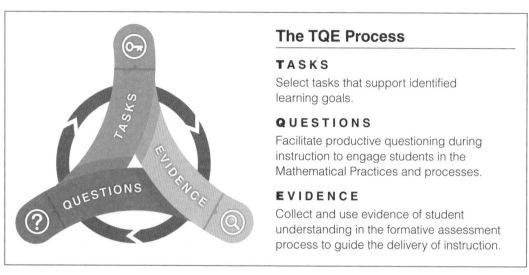

Source: Dixon, Nolan, & Adams, 2016, p. 4.

Figure I.1: The TQE process.

Throughout the book, we present a set of rubrics—the Instructional Quality Assessment (IQA) Mathematics Toolkit (appendix A, page 127, offers reproducible versions)—as a framework to focus reflections, conversations, feedback, and the planning and teaching of mathematics. The IQA rubrics provide a set of instructional practices and detailed descriptions of levels of quality within those practices. The Mathematical Tasks Framework and Levels of Cognitive Demand (Stein, Smith, Henningsen, & Silver, 2009) served as the foundation of the IQA. The set of IQA rubrics follow the progression of a mathematical task throughout a lesson as teachers engage students with the task, pose questions, orchestrate mathematical discussions, and collect evidence of students' learning. The IQA rubrics, and levels of cognitive demand within each rubric, form the core of this book and provide a way for teachers to focus on their instructional practice over time. Using the IQA rubrics, you will be able to identify instructional practices that support students' learning, areas for growth and improvement, and pathways for promoting that growth and improvement.

We encourage you to use *Making Sense of Mathematics for Teaching to Inform Instructional Quality* within a collaborative teacher team, which we define as at least two people with similar goals for improving the quality of mathematics instruction. In each chapter we encourage you to engage in activities, discuss your ideas about those activities together, read our analysis of the ideas in the activities, relate those ideas to the IQA rubrics, and apply the ideas and rubrics to your own mathematics classrooms.

Figure I.2: Play button icon.

Figure I.3: Task icon.

Throughout this book, we will use different icons to call your attention to various tasks to think about or perform. The *play button* icon, shown in figure I.2, indicates that an online video depicting a lesson is available for you to watch. You can find the videos either by scanning the adjacent QR code or by following the provided URL. (For a full list of videos and figures used in this book, see appendix E, page 143.)

The *task* icon, in figure I.3, highlights academic tasks to perform or problems featured in the videos. The tasks and lessons throughout the book represent tasks from a range of grade levels in elementary, middle, and high school. Regardless of the grade level or levels you teach, we have discovered that all teachers find value in exploring tasks and lessons across a variety of grade bands in order to illuminate the features of quality instruction and students' thinking.

Throughout the book, we ask you, as educators and collaborative teams, to focus on instructional quality by observing teaching. Sometimes we show this teaching through provided videos; at other times, we ask you to observe the teaching of members of your collaborative team. You may choose to accomplish this through live observations, by video recording the lessons, or by a combination of the two—whichever fits best within your individual contexts. Throughout the chapters, the IQA rubrics provide a helpful tool for teacher peers to both provide and receive feedback on the quality of your instructional practice—a highly important outcome allowing for more targeted and content-specific feedback than what is more frequently received from administrators, who may or may not have expertise in teaching mathematics (Darling-Hammond, 2014/2015).

Each chapter begins with introductory activities that engage you and your collaborative team with the key ideas in the chapter before we formally introduce the related IQA rubric or rubrics. Once you have an understanding of the rubric, we provide application activities to allow you and your team to practice using it and further reflect on your instruction. We encourage you to engage in the activities and discuss ideas with your collaborative team before moving on to the discussion following each activity. In each chapter, we provide resources that you may want to view or print as you complete the activities. These activities, materials, and videos are key to supporting your journey as you begin to reflect on mathematics instruction. To connect to practice, each chapter closes with a transition activity that applies the ideas in the chapter to the mathematics classroom and is then revisited in subsequent chapters. We close the book by providing you with the opportunity to use the entire IQA Toolkit to reflect on instruction and consider how to use IQA data to improve instruction.

We challenge you to reflect deeply as you explore one of the most influential characteristics related to student achievement—the quality of instruction.

PART 1
Connecting to the *T* in TQE: Tasks and Task Implementation

In this book, you will analyze and reflect on teaching mathematics at each stage of the TQE process using the Instructional Quality Assessment in Mathematics Toolkit rubrics. Part 1 connects to the *T* in the TQE process: "Tasks: Select tasks that support identified learning goals" (Dixon, Nolan, & Adams, 2016, p. 4). Implementing tasks that elicit thinking and reasoning can increase all students' access to high-quality mathematics. Throughout chapters 1 and 2, we highlight features of tasks and instruction in mathematics classrooms that promote access for all learners.

As you explore chapter 1, you will consider the impact of different types of *tasks* on students' learning of mathematics. In your work as a mathematics teacher (or with mathematics teachers), you have encountered many mathematical tasks—problems, exercises, homework sets, examples, activities, and so on—some of which have been interesting and challenging, and some of which have been routine and procedural. In this book, we use the term *mathematical task* to describe a problem or a set of problems that address a similar mathematical idea (Stein et al., 2009). A task can consist of a simple one-step problem, a complex multipart problem, or a series of related problems. Different types of tasks have different potential for engaging students in rigorous mathematics. We introduce the IQA Potential of the Task rubric in chapter 1 to provide a structure for analyzing the level and type of thinking a mathematical task might elicit from students.

Have you ever experienced a mathematics lesson in which you thought the task seemed simple, but surprisingly elicited much greater interest, thinking, and engagement than you anticipated? Conversely, have you experienced a mathematics lesson in which you anticipated the task to be interesting and engaging, but it somehow fell short of eliciting students' mathematical thinking and reasoning? The IQA Implementation of the Task rubric, introduced in chapter 2, will provide a structure for analyzing how instructional tasks play out during mathematics lessons.

CHAPTER 1

Potential of the Task

[There is] no decision teachers make that has a greater impact on students' opportunities to learn and on their perceptions about what mathematics is than the selection or creation of tasks with which the teacher engages students in studying mathematics.

—Glenda Lappan and Diane Briars

Why is it important to assess the cognitive potential of instructional tasks? First, the consistent use of high-level instructional tasks has been shown to enhance students' mathematical learning in elementary (Schoenfeld, 2002), middle (Cai et al., 2013), and high school mathematics classrooms (Grouws et al., 2013). Second, different types of tasks provide different types of opportunities for mathematical thinking and reasoning (Stein et al., 2009). Being aware of both the type of thinking a task can elicit and the type of access a task can give to all students can support you to align tasks with learning goals, and to ensure that students receive opportunities for thinking and reasoning. Finally, research has also shown that the level of the task sets the ceiling for the mathematical thinking, reasoning, and discussion that occurs throughout a lesson, and if a task does not request a representation, explanation, or justification, students typically do not produce or provide these things during a lesson (Boston & Wilhelm, 2015). Therefore, we find it critical for teachers seeking to improve their instructional practice to begin by considering the tasks and problems they are assigning in their classrooms and how these tasks may enable—or inhibit—student thinking.

What do you look for when selecting tasks? What makes a "good" instructional task?

In this chapter, you will explore why high-quality tasks are an essential first step in teaching mathematics for understanding. At the conclusion of this chapter, you will be able to answer the following questions.

- How do different types of tasks elicit different opportunities to learn mathematics?
- What types of tasks am I using to engage each and every student in learning mathematics?

Introductory Activities

Let's get started by thinking about different types of mathematical tasks. Activities 1.1 and 1.2 ask you and your collaborative team to solve a variety of mathematical tasks and consider the thinking and problem-solving strategies that each task might elicit.

Activity 1.1: Solving a Task

It is valuable to engage with tasks as learners prior to implementing them as teachers. Be sure to devote attention to this experience. Explore the task on your own before discussing your experience with others.

Engage

Solve the Leftover Pizza task in figure 1.1. Do not use any procedures or algorithms. Try to solve the task in more than one way, using diagrams or other representations, including in ways students might correctly or incorrectly solve this task.

> Douglas ordered 5 small pizzas during the great pizza sale. He ate ⅙ of one pizza and wants to freeze the remaining 4⅚ pizzas. Douglas decides to freeze the remaining pizza in serving-size bags. A serving of pizza is ⅔ of a pizza. How many servings can he make if he uses up all the pizza?

Source: Nolan, Dixon, Roy, & Andreasen, 2016.

Figure 1.1: The Leftover Pizza task (grade 6).

Respond to the following questions.

- What strategies and types of thinking can this task elicit?
- What are the main mathematical ideas that this task addresses?
- How do teachers typically present the mathematical ideas addressed in this task to students? What types of tasks do teachers typically use to present these mathematical ideas to students? What is different about this task?
- How might this task provide access for each and every learner?

Compare your work and ideas in your collaborative team before moving on to the activity 1.1 discussion.

Discuss

How do your responses compare with those in your collaborative team? What themes emerged during your discussion? In this section, we present ideas for you to consider.

What strategies and types of thinking can this task elicit?

The Leftover Pizza task is set in a context that is conceptually helpful for understanding the division of fractions. By thinking through the action in the problem, students can make sense of a situation that requires the division of fractions and solve the problem without needing to know a set procedure for dividing fractions. The context encourages the use of a drawing or manipulatives. Students are likely to draw circles or rectangles to model the pizzas, divide the pizzas into thirds or sixths, and create groups of ⅔ of a pizza. Students can also use pattern blocks to model the problem nicely, using the yellow hexagon as the whole, the blue rhombus as ⅓, and the green triangle as ⅙.

Students often determine that they can create seven whole servings of ⅔ of a pizza. The remaining piece of pizza elicits a dilemma and a common misconception in interpreting fraction division—the remaining piece is ⅙ of a pizza, but ¼ of a serving. Students often wrestle with determining if the answer is 7¼ or 7⅙ servings.

The task could be solved by applying a procedure for dividing fractions, but this would first require the student to make sense of the situation and realize (a) the need to divide 4⅚ by ⅔ and (b) what the answer of 7¼ *means* in the context of the problem. The ¼ refers to one of four parts of a serving of pizza, rather than ¼ of a whole pizza. The ⅙ refers to the part of the whole pizza remaining, rather than a part of the serving size.

What are the main mathematical ideas that this task addresses?

The Leftover Pizza task engages students in interpreting a contextual situation, dividing fractions, and interpreting the meaning of the quotient. While the main mathematics underlying the task is division of fractions, the task also provides opportunities for using diagrams or manipulatives, modeling a contextual situation, and making sense of the action in the problem and of the result. In this way, the task aligns with national standards, such as from the Common Core State Standards (CCSS) for mathematics: "Interpret and compute quotients of fractions, and solve word problems involving division of fractions by fractions, e.g., by using visual fraction models and equations to represent the problem" (National Governors Association Center for Best Practices [NGA] & Council of Chief State School Officers [CCSSO], 2010; 6.NS.A.1). The task also aligns with standards from the National Council of Teachers of Mathematics (NCTM, 2000): "Understand the meaning and effects of arithmetic operations with fractions, decimals, and integers" (p. 214).

How do teachers typically present the mathematical ideas addressed in this task to students? What types of tasks do teachers typically use to present these mathematical ideas to students? What is different about this task?

Educators often present fraction division as a rote procedure, modeling the process for students in example problems and accompanying this modeling with hints, such as "Remember to invert and multiply," or "keep-change-flip." Sometimes the examples are set in a context, but often students are provided a procedural solution to the examples and not encouraged to draw or model the situation or to make sense of the result. For example, students might be given the problem 4⅚ ÷ ⅔ along with several similar problems (for example, "Complete classwork examples 1–20") that could be solved by applying the same procedure to each problem. This set of problems (considered as one "task" according to our definition of a task as a set of problems that address the same mathematical idea) encourages students to apply a previously learned procedure, but does not support them to think or reason about division of fractions. The task directions and number of problems suggest that the focus of the task is on performing or practicing a procedure to produce an answer. The expected solution would look similar to:

$$4\frac{5}{6} \div \frac{2}{3} = \frac{29}{6} \div \frac{2}{3} = \frac{29}{6} \times \frac{3}{2} = \frac{29}{\cancel{6}2} \times \frac{\cancel{3}1}{2} = \frac{29}{4} = 7\frac{1}{4}$$

While 4⅚ ÷ ⅔ and the Leftover Pizza task both require the same mathematical operation and perhaps address similar content standards (dividing fractions), they provide much different opportunities for students' thinking and reasoning. The Leftover Pizza task engages students in interpreting, modeling, and making sense of a context that requires the division of fractions, the process of dividing fractions, and the meaning of the quotient. In this way, the Leftover Pizza task elicits the types of mathematical thinking identified in the Process Standards of NCTM's (2000) *Principles and Standards for School Mathematics*, such as *representations* and *connections*. It also engages students in the Mathematical Practices called for by the Common Core's Standards for Mathematical Practice, such as Mathematical Practice 1, "Make sense of problems and persevere in solving them," and Mathematical Practice 4, "Model with mathematics" (NGA & CCSSO, 2010).

How might this task provide access for each and every learner?

This task has many features that allow access for each and every student. It can be described as having a low threshold and a high ceiling (McClure, 2011). There are numerous ways to solve the pizza task, ensuring multiple entry points for all learners. Students can use fraction tiles, pattern blocks, or other manipulatives to model the situation and utilize those models when communicating how they solved the problem to their peers and teachers. The pizza task allows students access because they are able to model the serving sizes and determine that they can make at least seven servings. Then, using the models or through discussion with peers or the teacher, students can get to the ¼ of a serving that remains. By allowing students entry into the problem, teachers provide them with something to discuss with the class to further their understanding. If a student was just given the problem 4⅚ divided by ⅔ and did not have the means to perform the operation or solve the problem, he or she would not have access or the ability to solve the problem and would then be left out of the conversation regarding the solution. Multiple entry points and solution methods, as well as a meaningful context, allow students to interact with the task on at least some level so that when there is a discussion, all students have ideas to bring to the table and then can use those ideas to make sense of the mathematics and build a better understanding of the division problem.

In activity 1.2, you will continue to explore how different types of tasks provide different opportunities for students' thinking.

Activity 1.2: Considering Different Types of Tasks

It is valuable to engage with tasks as learners to make sense of what those tasks have to offer students. Be sure to devote attention to this experience. Explore the tasks on your own before engaging in the activity.

Engage

For activity 1.2, you may want to print figure 1.2 (page 13) and figure 1.3 (page 14) from this book or from the online resources (see **go.SolutionTree.com/mathematics**). Look over the tasks in figure 1.2 and use figure 1.3 to record your responses to the following questions.

- What is similar about the tasks in each column? How do the tasks change as you move up (or down) a column?
- What is similar about the tasks across each row? Identify phrases that characterize the nature of tasks in each row of the grid, and write these phrases on your recording sheet.

Compare your work and ideas in your collaborative team before moving on to the activity 1.2 discussion. Keep your recording sheet to use as your rubric in activity 1.3 (page 16).

Discuss

How do your responses compare with those in your collaborative team? What themes emerged during your discussion? In this section, we present ideas for you to consider.

What is similar about the tasks in each column? How do the tasks change as you move up (or down) a column?

The columns of the Benchmark Tasks grid each contain tasks that address related mathematical ideas. In column A, the tasks address division with remainders for students in grade 4. Tasks in column B relate to addition and subtraction of integers in grade 7. The tasks in column C all involve the area of trapezoids in grade 6. However, as you move up or down a column, the tasks provide different opportunities for students' thinking and reasoning about each mathematical topic. Research shows that attending to the level and type of thinking that a task can elicit from students is equally as important as considering the mathematical ideas in the task, and different types of tasks provide different opportunities for thinking and reasoning (Stein et al., 2009) and impact students' learning in different ways.

What is similar about the tasks across each row? Identify phrases that characterize the nature of tasks in each row of the grid, and write these phrases on your recording sheet.

Tasks across the rows of the Benchmark Tasks grid elicit similar types and levels of thinking from students. Tasks in row 1 mainly draw on students' memorized knowledge or recall of mathematics facts, rules, formulas, or vocabulary. Teachers have described tasks in this row as "You either know it or you don't"—nothing in the task helps students to learn what it is asking, and there is no procedure they can apply to determine an answer. Students only have access to the task if they are able to recall what it is asking. Taking notes would also belong in row 1 of this grid, as taking notes engages students in writing down or reproducing mathematics rather than doing any mathematical thinking on their own. Tasks such as these are appropriate when the goal for student learning is recall and memorization.

In row 2, students can solve the tasks by applying a procedure, computation, or algorithm. The goal of these tasks is to perform a procedure or computation and arrive at a correct answer. While students may use a conceptually based strategy to solve the task, nothing in the task requires or supports students to make sense of the mathematics or demonstrate their understanding of the mathematics. This denies access to the task if students are not fluent with the procedure or computation used to solve the problem. Successfully completing the task only requires that students perform a procedure and produce an answer.

In row 3, tasks may ask students to engage in problem solving, though they may also ask them to apply specific procedures or use specific representations. The main difference between tasks in row 2 and row 3 is that tasks in row 3 provide opportunities for mathematical connections, reasoning, and sense making. The questions, representations, and contexts in the task support students to develop an understanding of a mathematical concept or procedure or to engage in complex and non-algorithmic thinking. In completing the task, students actually learn mathematics.

Tasks in row 4 contain all of the features of tasks in row 3, with the added feature that the task directions explicitly require students to provide an explanation or justification. In addition to completing the mathematics necessary to solve the task, the task includes a prompt for students to reflect on, explain, or justify some aspect of their work.

	A (Grade 4 tasks)	B (Grade 7 tasks)	C (Grade 6 tasks)
4	Write word problems for 26 divided by 4 where: • The answer would need to be 7. • The answer would need to be 6. • You would need the exact answer. How are the three situations the same and how are they different? How is it possible to get a different answer to the same division problem?	What is the new net worth? Create two representations to make sense of the following scenarios. Describe how each of your representations models the mathematics in the situation. Rich began with a net worth of $700,000 but then made a very bad investment and lost $900,000. What is his new net worth? Notta began with a net worth of –$100,000 and then got a loan for $400,000 and spent the money to start a business writing screenplays. (a) What is her new net worth? (b) Notta then sold 6 different screenplays for $200,000 each. What is her new net worth? Ida began with a net worth of –$300,000. She took out 4 loans at $100,000 each to lend money to each of her 4 daughters to attend law school. What is her new net worth?	A trapezoid is shown below. Using any combination of rectangles, parallelograms, and triangles, determine a formula for the area of this trapezoid. Justify why your formula works.
3	Write a word problem for 26 divided by 4 that results in an answer of 7. Do not use the words *around*, *estimate*, or *about*.	What is the new net worth? Make sense of the following scenarios using a vertical number line. Rich began with a net worth of $700,000 but then made a very bad investment and lost $900,000. What is his new net worth? Notta began with a net worth of –$100,000 and then got a loan for $400,000 and spent the money to start a business writing screenplays. (a) What is her new net worth? (b) Notta then sold 6 different screenplays for $200,000 each. What is her new net worth? Ida began with a net worth of –$300,000. She took out 4 loans at $100,000 each to lend money to each of her 4 daughters to attend law school. What is her new net worth?	A trapezoid is shown below. Using any combination of rectangles, parallelograms, and triangles, determine a formula for the area of this trapezoid. Once you find one way, see if you can find another way.

Potential of the Task

2	Solve: 700,000 − 900,000 = −100,000 − 400,000 = −500,000 + 600,000 = Challenge: −300,000 + 4(−100,000) =	Using the formula, find the area of the trapezoid.

1	Copy in your notes the rules for determining the sign of the sum of two integers:
	
	  a. Positive + Positive → Positive
	  b. Negative + Negative → Negative
	  c. Positive + Negative or Negative + Positive →
	
	Sign of the integer with the larger absolute value
		What is the formula for the area of a trapezoid?

2	Divide: 26 ÷ 4 → _____ R _____ 17 ÷ 5 → _____ R _____ 43 ÷ 6 → _____ R _____	

1	For the following problems, underline the **dividend**, circle the **divisor**, put a square around the **quotient**, and put a triangle around the **remainder**.
	
	    6 R 2
	  4)26
	
	    3 R 2
	  5)17	

Source: Questions adapted from Dixon, Nolan, Adams, Tobias, & Barmoha, 2016, p. 60; Nolan, Dixon, Roy, & Andreasen, 2016, pp. 36, 105.

Figure 1.2: Benchmark Tasks grid.

Visit go.SolutionTree.com/mathematics for a free reproducible version of this figure.

Row	Comments
4	
3	
2	
1	

Figure 1.3: Benchmark Tasks recording sheet.

Visit **go.SolutionTree.com/mathematics** *for a free reproducible version of this figure.*

Students can use prior knowledge to help solve the problems in rows 3 and 4, which allows more access to these types of problems. For example, in problem 2C, if students do not know the formula for a trapezoid, they most likely would be unsuccessful in solving and would not attempt the problem. However, in problems 3C and 4C, students could use their knowledge of other formulas to break up the trapezoid into other shapes, find the area of those shapes, and then find the area of the figure. While this may not be the most efficient strategy, it allows access to the problem and students can find a solution. Then, through discussion with peers, students can connect their solution to others and engage in the lesson. It is interesting that tasks that teachers often consider to be more difficult actually provide more access to students. Additionally, if students are not able to complete the mathematics necessary to solve the task, the explanations and justifications the students provide can assist the teacher in diagnosing gaps in students' understanding so that the teacher can then address those gaps. (Note: In this book, we use the terms *demanding* and *challenging* to mean stimulating and thought provoking, rather than difficult. A difficult task—for example, multidigit long division—may be difficult but not necessarily cognitively challenging.)

How did your responses on the Benchmark Tasks recording sheet compare to these descriptions? Take a moment to consider or discuss any new ideas introduced in this section using your recording sheet from activity 1.2. Then, proceed to the following section, The IQA Potential of the Task rubric, where we present a framework from the IQA Toolkit to assist you and your collaborative team in assessing the potential cognitive demand of mathematical tasks.

The IQA Potential of the Task Rubric

The level and type of thinking required by a task is referred to as the *level of cognitive demand* (Stein et al., 2009). Teachers can use the IQA Potential of the Task rubric (figure 1.4) to rate the level of cognitive demand a task can elicit from students. In activity 1.2, you identified examples and features of tasks at different levels of cognitive demand. The identification of these task features will help you to rate a task using the Potential of the Task rubric, with which you will consider the level of thinking required for students to successfully complete the task as initially written, without any changes, extensions, resources, or other expectations that may arise as students work on the task.

IQA Potential of the Task Rubric	
4	The task has the potential to engage students in complex thinking or in creating meaning for mathematical concepts, procedures, or relationships. The task *must explicitly prompt* for evidence of students' reasoning and understanding. For example, the task may require students to: • Solve a genuine, challenging problem for which students' reasoning is evident in their work on the task • Develop an explanation for why formulas or procedures work • Identify patterns and form and justify generalizations based on these patterns • Make conjectures and support conclusions with mathematical evidence • Make explicit connections between representations, strategies, or mathematical concepts and procedures • Follow a prescribed procedure in order to explain or illustrate a mathematical concept, process, or relationship
3	The task has the potential to engage students in complex thinking or in creating meaning for mathematical concepts, procedures, or relationships. However, the task does not warrant a level 4 rating because it does not explicitly prompt for evidence of students' reasoning and understanding. For example, students may be asked to: • Engage in problem solving, but for a task that provides minimal cognitive challenge (for example, a problem that is easy to solve) • Explore why formulas or procedures work, but not to provide an explanation • Identify patterns, but not to explain generalizations or provide justification • Make conjectures, but not to provide mathematical evidence or explanations to support conclusions • Use multiple strategies or representations, but not to develop connections between them • Follow a prescribed procedure to make sense of a mathematical concept, process, or relationship, but not to explain or illustrate the underlying mathematical ideas or relationships
2	The potential of the task is limited to engaging students in using a procedure that is either specifically called for, or its use is evident based on prior instruction, experience, or placement of the task. • There is little ambiguity about what needs to be done and how to do it. • The task does not require students to make connections to the concepts or meaning underlying the procedure they are using. • The focus of the task appears to be on producing correct answers rather than developing mathematical understanding (for example, applying a specific problem-solving strategy or practicing a computational algorithm).
1	The potential of the task is limited to engaging students in memorizing; note taking; or reproducing facts, rules, formulas, or definitions. The task does not require students to make connections to the concepts or meanings that underlie the facts, rules, formulas, or definitions they are memorizing or using.

Source: Adapted from Boston, 2017.

Figure 1.4: IQA Potential of the Task rubric.

The IQA Potential of the Task rubric is intended to align with our previous ideas about tasks in rows 1 through 4 of the Benchmark Tasks recording sheet (page 14)—which we hereafter refer to as *levels* in the IQA Toolkit—and to provide additional detail and support for rating tasks. Look back through your task ratings and rationales for activity 1.2 and consider the following questions.

- In what ways does the Potential of the Task rubric appear to be *consistent with* ideas on your recording sheet from activity 1.2? What words or phrases on the rubric do you find helpful?

- In what ways does the Potential of the Task rubric appear to be *inconsistent with* or *different from* ideas on your recording sheet from activity 1.2? In other words, what characteristics of tasks did you identify that are not represented in, different from, or in contrast with the rubric?

Talk about the consistencies and inconsistencies with your collaborative team before moving on to the Application Activities in the following section. While several features of tasks may be important, this framework captures differences in tasks that have been shown to generate differences in students' mathematical learning (Grouws et al., 2013; Stein & Lane, 1996). The way we categorize tasks according to cognitive demand frames many ideas throughout this book, so it is important to spend the time now to resolve differences with ideas in the rubric and within your collaborative team. These activities will assist you further in using the IQA Potential of the Task rubric (figure 1.4) and assessing your current instructional practices.

Application Activities

The following activities will help you become familiar with the IQA Potential of the Task rubric as you practice rating and adapting mathematical tasks.

Activity 1.3: Rating Mathematical Tasks Using the Potential of the Task Rubric

It is valuable to engage with tasks as learners to make sense of what those tasks have to offer students. Be sure to devote attention to this experience. Explore the tasks on your own before engaging in the activity.

Engage

For activity 1.3, you may want to print figure 1.5 from this book or the online resources. Note that we have provided grades or grade bands for each task. Because specific mathematics standards may vary from state to state, assume the task is appropriate for the grade level and students for which it is being used.

As you complete the task, consider the following directions.

- Rate each task in figure 1.5 from level 1 to level 4 using the Potential of the Task rubric and provide a reason for the level you selected. Determine the ways each task provides access to each and every student.

- Discuss your ratings and ideas with your collaborative team before moving on to the activity 1.3 discussion.

Potential of the Task

Water Fountain Task (Algebra)

The spray from a water fountain is modeled by the quadratic functions $f(x) = -(x-5)^2 + 25$ and $f(x) = -x^2 + 10x$.

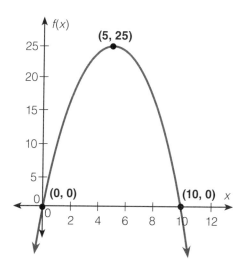

Which form of the quadratic equation would you use to determine the following information if the graph was not provided? Why would you choose that form?

1. The maximum height reached by the spray
2. The distance between the starting point and the ending point of the spray

Multistep Equations Task (Grade 7)

In problems 1–8, solve the following equations. Show your work!

1. $x - 17 = 31 - 2x$
2. $(n - 31) = 108$
3. $97 = w - 15 + 3w$
4. $17 + b = 37 - 3b$
5. $6(2a - 1) = 54$
6. $4d = 3d - (6d - 21)$
7. $5y - (2y + 3) = -33$
8. $10 - 3z = 5z - 6$

Division Story Problems Task (Grade 5)

Use division to solve the following word problems. Write the equation you used to solve the problem. Write your answer in a complete sentence.

1. There are 124 craft sticks in a package. You have one package of craft sticks. It takes 4 craft sticks to make a picture frame. How many picture frames can you make?
2. The deli has 175 slices of turkey. Each sandwich uses 5 slices of turkey. How many sandwiches can the deli make?
3. The Office Supply Store receives bulk boxes of 200 pencils. They make smaller packages of 8 pencils each to sell. How many packages of pencils can they make?

Science Quiz Task (Grade 7)

Mr. Richards and Ms. Chutto decide to compare the grades in their two science classes on the last quiz.

The grades in Mr. Richards's class on the twenty-point quiz were:

15	16	15	18	17	11	16	14	12	15	7	14
16	8	7	17	6	8	15	13	14	11	14	

The grades in Ms. Chutto's class on the twenty-point quiz were:

13	12	9	10	9	20	8	16	10	11	17	12
13	12	14	12	11	15	15	18	15	19	14	

Whose class did better on the quiz? How do you know?

Source: Questions adapted from Dixon, Nolan, Adams, Tobias, & Barmoha, 2016, p. 84; Nolan, Dixon, Roy, & Andreasen, 2016, pp. 128, 133; Nolan, Dixon, Safi, & Haciomeroglu, 2016, p. 46.

Figure 1.5: Tasks for activity 1.3.

continued →

Shapes Pattern Task (Grade 2)	**Swimming Pool Deck Task (Grade 6)**
Imagine the following pattern continues:  • What shapes will be at steps 9, 10, 11, and 12? • What will the shape be at step 17? • What will the shape be at step 39? • What is special about the step numbers that have a sun? • Make some conjectures about the shapes that will be at really large step numbers, even step numbers, and odd step numbers. List any other things you notice about the pattern.	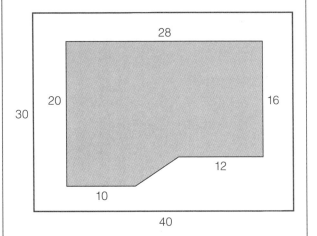 Erik is building a swimming pool in his backyard. He wants to tile the deck around the pool. The picture above shows the dimensions (in feet) of the deck and pool. How much area will need to be tiled?
Properties of Multiplication Task (Grade 5)	**Fraction Pizza Task (Grade 3)**
Indicate whether each statement is an example of the commutative property, associative property, identity property, or zero property of multiplication: a. $(3 \times 5) \times 4 = 3 \times (5 \times 4)$ b. $13 \times 0 = 0$ c. $17 \times 3 = 3 \times 17$ d. $1 \times 13 = 13$	Problem 1: Marc and Larry each bought a small cheese pizza for lunch but neither of them was very hungry. Marc ate $1/6$ of his pizza and Larry ate $1/5$ of his pizza. Who ate more pizza? Problem 2: Riley and Paige each bought a small cheese pizza for lunch. They were both very hungry. Riley ate $5/6$ of her pizza and Paige ate $6/7$ of her pizza. Who ate more pizza?

Visit go.SolutionTree.com/mathematics for a free reproducible version of this figure.

Discuss

Rate each task in figure 1.5 from level 1 to level 4 using the Potential of the Task rubric and provide a reason for the level you selected. Determine the ways each task provides access to all students.

Level 4 refers to tasks that promote meaning, sense making, connections between representations, or problem solving and explicitly require explanations or justifications. We rated the following two tasks at level 4.

- **Water Fountain task:** This task provides an opportunity for students to make connections between representations. It provides a context, a graph, and two forms of a quadratic function. The task asks students to consider which of the symbolic representations would be most useful for answering different questions about the water fountain. It explicitly prompts them to explain their choices. Students have access to the task because they are able to visualize the representation of the graph and relate the functions to the graph.

- **Science Quiz task:** This task requires students to determine how to compare the data. Students must make sense of the type of data, the distribution of scores, and what these both suggest about appropriate representations to model and compare the data. Students are solving a genuine problem and developing an explanation for why their choices make sense. The task explicitly prompts students to explain how they know which class did better on the quiz. This task allows access because there are multiple ways to compare the sets of data and make an argument using mean, median, mode, range, dot plots, and box plots. Students could argue for either class depending on what measure of center they choose or which type of graph they create and are not confined to using one particular procedure.

Level 3 refers to tasks that promote meaning, sense making, connections between representations, or problem solving but do not explicitly require explanations or justifications. We rated the following three tasks at level 3.

- **Shapes Pattern task:** Identifying patterns and forming conjectures provide opportunities for thinking and reasoning as well as recognizing and using structure. While the task asks students to make conjectures, the task does not prompt students to provide mathematical evidence for those conjectures. This task can engage students in thinking about important mathematics at grade 2 (for example, multiples or division with remainders). This task allows access because most students can identify a pattern and then engage in a conversation with peers around how to justify and generalize the pattern. Including a prompt such as "How do you know?" or "Determine whether your conjecture is always true" would increase the task to a level 4.

- **Swimming Pool Deck task:** This task would be rated as a level 3 because it provides a context and opportunity for students to make sense of area, but it is not a level 4 because it does not ask students to form a generalization or justify their solutions. The shape is nonstandard, and students cannot just apply an area formula and obtain an answer. There are multiple ways to find the area of the deck and the task suggests no specific pathway to the students. The task provides access because students who cannot recall the formula for finding the area of a trapezoid can use other area formulas by decomposing the shape into other shapes for which they know the formulas.

- **Fraction Pizza task:** The task has the potential to engage students in complex thinking and creating meaning about the relative size of fractions. The task provides a context in which students can compare the relative size of unit fractions and fractions one part away from a whole. While students could use a procedure to compare fractions (for example, common denominators), the task provides a context to support students to reason about the relative size of fractions close to 0 and close to 1, even if they did not know the procedure. The opportunities inherent in this task to solve it in different ways increase its access to more students. Note that the task requires no explanation, hence it is not a level 4.

Level 2 refers to tasks that require procedures, computation, or algorithms without connection to meaning and understanding. We rated the following two tasks at level 2.

- **Multistep Equations task:** The potential of the task is limited to students performing a procedure or series of procedures to solve multistep equations. The number of problems in the set suggests that the task requires students to apply procedures quickly and efficiently. Solving equations is an important and useful algebra skill, and this particular task provides the opportunity for students to practice and demonstrate their ability to perform previously learned procedures for solving equations. For this reason, the task is a level 2. The task does not support students to develop an understanding of the underlying mathematical concepts (for example, the property of equality). The prompt to "show your work" does not require students to explain their thinking and reasoning, but simply to show the steps in the procedure. This task does not allow access if students do not already have a set procedure for how to solve multistep equations.

- **Division Story Problems task:** The potential of the task is limited to engaging students in a procedure that the task specifically calls for. The directions tell students exactly what operation to use; there is little ambiguity about what to do, removing students' opportunities for thinking and sense making. The main mathematical activity left for students is dividing the first number in the problem by the second number in the problem. Even though this set of problems is set in a context (for example, "word problems" or "story problems"), this task is a level 2. Each problem follows a very similar format, and students can apply a procedure given to them in the directions of the task (division) and obtain an answer without considering the action in the problem or making sense of the situation. While the context of each problem supports an understanding of the equal groups or measurement model of division, the directions instruct students to use division at the outset of the problem, thus minimizing their opportunity to think through an appropriate model and operation themselves. Note that removing the directions and varying the format of each situation would make the task a level 3, as students would then need to make sense of the situations and select an appropriate model. Then, also including the prompt "Explain how you know your strategy and solution make sense" would raise the task to a level 4. While this task rates a level 2, it still provides good access to students because it is situated within contexts that allow students to solve the problems in multiple ways, including drawing pictures or models of each situation.

Level 1 refers to tasks that elicit recall and memorization. We rated the following task at level 1.

- **Properties of Multiplication task:** The potential of the task is limited to engaging students in recalling memorized knowledge of the properties of multiplication. Nothing in the task helps the students learn about the properties; they are simply asked to name the property displayed in each example. If students do not know each property, they are not able to access this task. It is difficult to modify tasks that are level 1 to increase access for more students without altering the mathematical goal of the task.

Activity 1.4 provides an additional opportunity to use the Potential of the Task rubric to rate and adapt the levels of tasks.

Activity 1.4: Using the IQA Potential of the Task Rubric to Rate and Adapt Tasks

It is valuable to engage with tasks as learners to make sense of what those tasks have to offer students. Be sure to devote attention to this experience. Explore the tasks on your own before engaging in the activity.

Engage

For activity 1.4, you may want to print figure 1.6 (page 22) from this book or the online resources. The tasks in figure 1.6 are examples of tasks at levels 1 through 3.

- Provide a rationale for each task level using the IQA Potential of the Task rubric.

- Consider how to adapt each task to increase the cognitive demand. Use ideas from the Potential of the Task rubric to make small changes to each task to provide greater opportunities for students to provide their thinking and reasoning while still addressing the same mathematical content.

- Before moving on to the activity 1.4 discussion, discuss your rationales and task adaptations with your collaborative team. Include in your discussions how your task adaptations might increase the potential for access to the task by more learners. Compare your ideas with the rationales and suggestions for adaptations in appendix B (page 137).

Discuss

How do your responses compare with those in your collaborative team? What themes emerged during your discussion? In this section, we present ideas for you to consider.

Provide a rationale for each task level using the IQA Potential of the Task rubric.

Rationales for the levels of tasks in figure 1.6 appear in appendix B (page 137). Were your rationales consistent with the ones provided? Were there any ratings you questioned? In the following description of adaptations, we provide additional detail regarding the rating of each task.

Consider how to adapt each task to increase the cognitive demand.

In this activity, we suggested using ideas from the Potential of the Task rubric to make small changes to adapt the tasks in figure 1.6 to provide greater opportunities for students' thinking and reasoning, while still addressing the same mathematical content. Here, we describe three types of changes that would increase the tasks to a level 3 or 4.

1. **Level 3: Number Pairs That Make 10**—The Number Pairs task has the potential to engage kindergarten students in creating meaning for how to generate sums to ten. It is important for students at this age to know how to compose and decompose numbers, especially with tens. Because we would not yet expect kindergartners to have memorized the number facts that sum to ten, this task allows students to explore different ways to make ten. There are many ways students could think through the problem and model their ideas and strategies, which is an important feature in providing access to all students. The Number Pairs task does not rate a level 4 because there is no explicit prompt for an explanation or justification. This

22 MAKING SENSE OF MATHEMATICS FOR TEACHING TO INFORM INSTRUCTIONAL QUALITY

Task	Rationale	Ways to Adapt the Task to Increase the Cognitive Demand
Level 3 Number Pairs That Make 10 (Grade K) Two-sided chips (red and yellow) are available. Jasmine has 10 marbles. Some of them are red and the rest are yellow. How many marbles could be red and how many marbles could be yellow?		
Level 2 Adding Fractions With Unlike Denominators (Grade 5) In problems 1–3, find a common denominator and add the fractions: 1) $\frac{3}{4} + \frac{1}{20} =$ 2) $\frac{2}{3} + \frac{3}{5} =$ 3) $\frac{5}{12} + \frac{1}{6} =$		
Level 1 Angles (Grade 7) Lines AB and CF are parallel. Name pairs of angles that are: a. Vertical angles b. Supplementary angles c. Alternate interior angles d. Corresponding angles 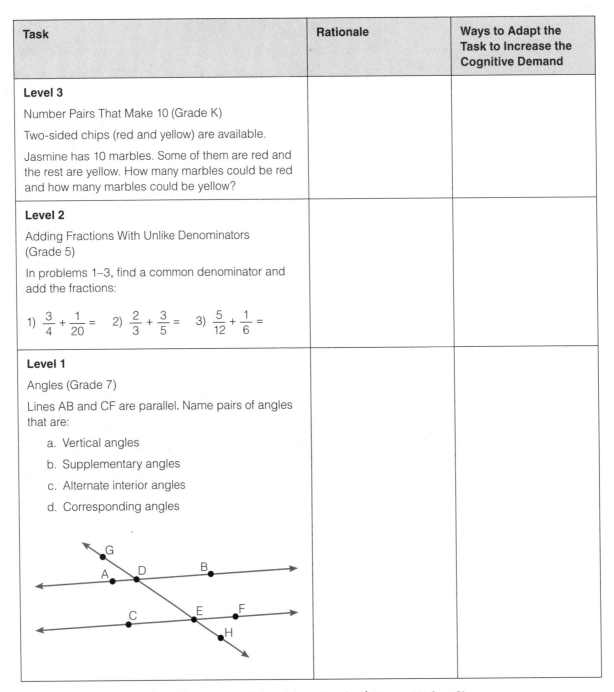		

Source: Level 3 question adapted from Dixon, Nolan, Adams, Brooks, & Howse, 2016, p. 65.

Figure 1.6: Tasks for activity 1.4.

Visit **go.SolutionTree.com/mathematics** for a free reproducible version of this figure.

task would provide greater opportunities for thinking and reasoning than traditional tasks that ask students only for answers such as:

$$5 + 5 = \underline{\qquad} \qquad 8 + 2 = \underline{\qquad} \qquad 3 + 7 = \underline{\qquad}$$

Note, however, that with older students who have the number facts memorized, we would characterize the Number Pairs task as level 1 (memorization).

In general, level 3 tasks provide opportunities for thinking, reasoning, and sense making. Added prompts for students to explain their thinking, compare strategies, reflect on their strategy choice, or justify their conjectures or generalizations are examples of how to raise the task to a level 4. Asking students to find all possible solutions, and to explain how they know they have found them all, can engage students in analyzing patterns and making generalizations. Alternatively, asking students for two different ways to solve the problem or to find more than one solution, and prompting students to explain, compare, or relate the different solutions, also increases the cognitive demand. In the Number Pairs task, asking students to explain why more than one number pair works would more deeply engage students in decomposing and recomposing numbers and explaining their reasoning. Finally, requiring students to create a representation and explain something about the representation can also increase cognitive demand.

2. **Level 2: Adding Fractions With Unlike Denominators**—The Adding Fractions task is a typical procedural task. There are numerous procedures for every grade level and mathematical topic that we could substitute in place of "adding fractions with unlike denominators" (for example, multiplying or dividing multidigit numbers, cross-multiplying, applying the Pythagorean theorem, or factoring). Such tasks provide opportunities for students to practice or demonstrate a previously learned procedure. While practice or mastery of certain mathematical procedures is often useful and even necessary, as teachers we want to be aware that engaging in procedural tasks only promotes practice and rote mastery and does not promote understanding and sense making. For example, one can know and perform the procedure for dividing multidigit numbers but not be able to explain why you "bring the number down" or know when division applies to a contextual situation. For these reasons, more conceptual tasks (levels 3 and 4) align with goals and standards when students are beginning to develop an understanding of the mathematical topic, and procedural tasks may align better with goals and standards at the end of students' learning trajectory of a particular mathematical topic, once they have developed their understanding.

   To increase the cognitive demand of a procedural task, use a context or representation that supports students to make sense of the operation and provides the need to develop a new strategy. To increase access, allow students to use multiple strategies or manipulatives to engage with the task. Consider the following task from *Making Sense of Mathematics for Teaching Grades 3–5* (Dixon, Nolan, Adams, Tobias, & Barmoha, 2016):

   > Brandon is sharing four cookies equally between himself and his four friends. Brandon wants to start by giving each person the largest intact piece of cookie possible so each person receives the same size piece of cookie to start. How might Brandon divide the cookies? (p. 73)

While fifth-grade students might easily determine that four cookies shared among five people is $4/5$ of a cookie per person, requiring the largest intact piece of cookie to be shared equally first provides a context for adding fractions with unlike denominators, such as $1/2 + 1/4 + 1/20$. Many other scenarios and contexts can support the need for students to make sense of adding equal-sized fractional parts.

Another approach to increasing the cognitive demand of a task is to ask students to develop a new procedure based on prior knowledge before teaching the procedure to students. In this case, knowledge of equivalent fractions and adding fractions with like denominators is all students need to figure out how to add fractions with unlike denominators. Similarly, removing structure or directions that prescribe a strategy or direct students how to solve the task will also raise the task level. Beyond procedural tasks, this adaptation also applies to word problems or story problems at all grade levels and for any mathematical procedure. For example, if high school students have just learned the distance formula, and then are given word problems with the directions "Use the distance formula to solve the following problems," the word problems no longer require thinking and reasoning, but only the application of a previously learned, prescribed procedure. This lowers the cognitive demand of the task.

3. **Level 1: Angles**—Similar to the idea that level 2 (procedural) tasks are appropriate when the goal for students' learning is practice or mastery of procedures, level 1 tasks are appropriate when the goal for student learning is recall and memorization. To encourage greater thinking and reasoning, allow students to discover relationships *before* providing the vocabulary, definitions, properties, postulates, or theorems. For example, before providing the definition and properties of vertical angles, have students measure a variety of vertical angles, conjecture that vertical angles are congruent (which also provides an opportunity to discuss measurement error), and then justify their conjecture using prior knowledge (for example, supplementary angles and the transitive property).

Although the Potential of the Task rubric provides a comprehensive framework for rating and adapting mathematical tasks, certain factors may affect how you rate or select tasks for classroom use. We discuss these issues in the following section.

Considerations When Rating Tasks

The awareness of different task levels and the ability to rate the level of tasks can equip teachers to be knowledgeable and critical consumers of published and online resources for the mathematics classroom. Published curricular materials often contain tasks at a variety of levels, and small changes to adapt the tasks in ways such as those identified in activity 1.4 can go a long way toward increasing students' opportunities for thinking and reasoning. Instructional materials featured in online sites for teachers are frequently divided between resources that promote procedural practice and nonmathematical activity (for example, when the main activity is craft based rather than mathematical) and resources that provide ideas for conceptually based lessons, and teachers often have to adapt these resources for use in their own classrooms.

As you begin to assess and rate tasks, there are several issues to consider in order to achieve both successful implementation in your classroom and enhanced thinking and understanding among your students. In this section, we will discuss the practical and conceptual issues that may stem from defining the task itself, considering the implications of higher-level thinking in practice, and aligning tasks with learning goals and standards.

Defining the Task

Sometimes, identifying the task in curricular materials or other resources is not straightforward. In this book, we consider the task to be the mathematical problem or set of problems presented for students to do during a lesson or instructional activity. Tasks in curricular materials or as presented during a lesson may contain several parts. For example, each cell of figure 1.2 (page 13) is considered to be one task. The multiple parts of a task receive one collective rating according to the highest level of cognitive demand of any of the parts. For example, if a task (as presented on a handout or in verbal directions) includes vocabulary recall, a problem-solving activity, and an explanation, we would rate it a level 4. A task that supports students to develop or generalize a procedure and then spend time practicing that procedure would be considered a level 3.

Some curricular materials will clearly identify a mathematical problem or set of problems for students to engage with during the lesson. For other materials, you may need to identify what the task is asking students to do mathematically. Sometimes, the teacher's manual is necessary to understand exactly what students are being asked to do. This is particularly true for primary grades, in which the teacher often presents the task and directions verbally to align with students' reading levels. When rating tasks in curricular materials or other resources, consider them as they appear in print. Any directions, manipulatives, representations, or resources indicated by the print materials, including teachers' manuals, are part of the task.

When using the Potential of the Task rubric (page 15), consider any directions (via textbook, teacher's edition, handout, whiteboard, or screen) or resources provided to students. Most of the directions will occur before students begin their work on the task. However, the teacher may choose to give students part of a task, allowing them time to explore, and only then provide later parts of the task and additional time for students to continue working or developing their explanations. We would consider any mathematical problems that teachers ask students to do during the lesson as part of the task, even if the task directions are presented in parts throughout the lesson. In later chapters, we explore how a task unfolds throughout a lesson as students work on the task and engage in mathematical discussion, and so it is helpful to consider a lesson as having one main instructional task.

Considering Implications of Higher-Level Thinking

When discussing the activities so far in this chapter, you and your collaborative team may have occasionally determined that the level of the task depends on the grade level or prior knowledge of the students. It is always important to consider how the students' prior knowledge may impact the cognitive demand of a task. For the activities in the chapter, we instructed you to assume that the task was appropriate for a given group of students. In your school or classroom, you would be familiar with the grade level, standards, and students for a given task. If students have solved a series of very similar patterning tasks, problem-solving tasks, or other level 3 or 4 tasks, subsequent tasks in the series would not elicit the same level and type of thinking as the first. The task would likely become procedural (level 2), with students following a template provided by completing the first few tasks, even though the first task in the series would have been a level 3 or 4. Similarly, if students already know the properties, procedure, or definitions at the heart of a discovery task, there would not be anything for them to discover. For example, the adapted Angles task can help students discover relationships between angles *only if* students do not already know those relationships.

Additionally, be aware of the implications of the wording of the task and how this can impact the work students produce. While a task may ask students to *explain how* or *show your steps*, there is a difference between explaining a procedure and explaining your thinking. For example, if a task requires students to solve fraction division problems such as ¾ ÷ ¼ using the traditional algorithm, asking students to explain might generate responses such as, "I used invert and multiply and computed ¾ × ⁴⁄₁." While this explanation of how students solved the problem indicates knowledge of an algorithm, it does not indicate conceptual understanding, reasoning, or sense making. Adding a prompt to explain onto a procedural task does not raise the cognitive demand; the task itself must first elicit some thinking, reasoning, problem solving, or understanding for the student to have something worth explaining.

Aligning With Learning Goals

While we categorize procedural tasks at a level 2, note that there are many important mathematical procedures that students should be able to apply fluently and with automation after having established an appropriate level of conceptual understanding (NCTM, 2014). There are also appropriate occasions when you would use or assign a level 2 task to provide students the opportunity to practice or demonstrate their ability to perform a procedure, or a level 1 task when the goal is memorizing rules, properties, or definitions. Task levels (as well as mathematical content) should align with the goals for students' learning.

Task levels should also align with students' learning progression for a particular mathematical idea at a particular grade level. For example, rigorous state standards suggest providing opportunities consistent with level 3 or 4 tasks while students are in the process of learning to make sense of multiplication in grades 2 through 4, as expressed in the grade 4 Common Core standard:

> Multiply a whole number of up to four digits by a one-digit whole number, and multiply two two-digit numbers, using strategies based on place value and the properties of operations. Illustrate and explain the calculation by using equations, rectangular arrays, and/or area models. (NGA & CCSSO, 2010; 4.NBT.B.5)

By grade 5, we expect students to demonstrate mastery of the standard algorithm for multidigit multiplication, which would be supported by engaging in tasks at level 2 that provide practice and promote automation and mastery.

Note that in the progression of learning multiplication from grades 2 to 5, opportunities for students to understand, unpack, and develop strategies and procedures, using problems, contexts, and representations that make sense, *precede* the memorization of multiplication facts and mastery of standard algorithms. Too often, students' learning of a mathematical topic or procedure begins with the teacher telling or showing students everything they need to know, modeling procedures, or providing definitions, while students' mathematical activity is limited to practicing procedures (level 2) or taking notes and memorizing (level 1). Typically, teachers wait until students have mastered procedures or memorized the appropriate facts or properties before providing opportunities for real-world applications or problem solving. After all, how would students know how to solve problems if they were not shown how to solve them first? On the contrary, students can develop mathematical procedures given contextual problems prior to any direct instruction or modeling (Carpenter, Fennema, Franke, Levi, & Empson, 2014). Students can discover many mathematical properties through investigation, such as the congruent angles formed when parallel lines are cut by a transversal (figure 1.6, page 22), the formula for area of a trapezoid (figure 1.2, page 13), or the rules for adding integers (figure 1.2, page 13). The *Making Sense of Mathematics for Teaching* series provides many suggestions for supporting students' learning of specific mathematical topics at each grade band.

Summary

It is important to rate the level of instructional tasks to be aware of the type of thinking and access a task can provide for each and every student. A task at level 1 or 2 does not provide much space for discussion, as the focus is on the correctness of memorized knowledge or rote procedures. Additionally, a task at level 1 or 2 often does not provide access for students unless they know the mathematics to be recalled or the specific procedure requested. A level 3 or 4 task is often necessary to support quality mathematical discourse and teacher questioning, as we will discuss in upcoming chapters. We provide additional support for rating tasks in appendix D (page 141).

Even though reform efforts call for mathematics learning for each and every student (NCTM, 2014), learners who struggle in mathematics or who have special education placements often have less access to demanding mathematics (Weiss, Pasley, Smith, Banilower, & Heck, 2003). To successfully include all learners in the mathematics classroom, we need to design instruction that is accessible to all.

Chapter 1 Transition Activity: Moving From Tasks to Implementation

Before moving on to chapter 2, engage in the transition activity with your collaborative team. The transition activity will enable you to build on ideas about tasks from chapter 1 to begin to explore implementation in chapter 2.

- Select a chapter, unit, or any set of two to three consecutive lessons in the mathematics curriculum materials you use in your school or classroom. Rate the tasks that appear in a set of lessons over two to three days of instruction.
    - What opportunities would students have to engage in thinking and reasoning?
    - What is the balance of levels across the lessons?
- Identify a task at level 3 or 4 to use as the main instructional task to teach a mathematics lesson. Indicate what features make the task a level 3 or 4.
    - What thinking, reasoning, or sense making would the task potentially elicit from students?

- What products or processes would serve as evidence that students actually engaged in this thinking, reasoning, or sense making?
- How does the task provide access for all students?

- Implement the task in your class. Collect sets of student work (at least four samples). Select samples that show a variety of strategies, thinking, and reasoning.
- Analyze students' responses. Did students actually engage in or produce the level and type of thinking you identified when considering the potential of the task?

Save the sets of student work, your ratings, and notes or any written reflections from the transition activity, as you will refer to them in chapter 2.

CHAPTER 2
Implementation of the Task

To ensure that students have the opportunity to engage in high-level thinking, teachers must regularly select and implement tasks that promote reasoning and problem solving.

—National Council of Teachers of Mathematics

In this chapter, you will explore how teachers can implement mathematical tasks during mathematics lessons in ways that support or possibly diminish students' opportunities to engage in thinking and reasoning. By the end of the chapter, you will be able to answer the following questions.

- What happens when teachers enact high-level tasks in the classroom? How do teachers maintain (or limit) opportunities for thinking and reasoning during the lesson?
- What teacher actions seem to support (or diminish) students' opportunities for thinking and reasoning as well as provide (or take away) access for students to engage in the task?
- What does student work indicate about students' engagement in thinking and reasoning during the lesson and their level of access to the lesson?
- What does students' work indicate about instruction and teaching during the lesson?

Introductory Activities

Let's get started by analyzing mathematics lessons and observing how the implementation of a task can affect student learning. In activity 2.1, we ask you and your collaborative team to observe two mathematics lessons that use the same task, but in noticeably different ways. You will then consider the thinking and problem-solving strategies that each teacher's implementation elicited.

Activity 2.1: Comparing Two Mathematics Lessons

Before engaging in this activity, consider what opportunities for mathematical learning the Leftover Pizza task could offer students and how you might implement the task in the classroom.

Engage

In chapter 1, activity 1.1 (page 7), you considered the level of thinking and reasoning that could potentially be provided by the Leftover Pizza task for students in grade 6. In activity 2.1, you will explore how the task plays out in two different lessons.

- Revisit the Leftover Pizza task (see figure 1.1, page 8).
    - Rate this task using the Potential of the Task rubric (see figure 1.4, page 15).

- Be sure to include your rationale—what characteristics or features influenced your rating of the task? Discuss the Potential of the Task level with your collaborative team before moving on.
- Watch the video of the Leftover Pizza lesson version 1.
  - Take notes regarding what you notice in the Leftover Pizza lesson version 1. Write down timestamps from the lesson so that you can readily refer to what you noticed when you discuss the lesson within your collaborative team. Throughout the book, we use timestamps to identify particular features of video lessons.
  - How would you describe students' thinking and reasoning during the lesson? Include examples and timestamps.
- Read the transcript of the Leftover Pizza lesson version 2 (see figure 2.1).
  - What do you notice about the Leftover Pizza lesson version 2?
  - How would you describe students' thinking and reasoning in the narrative?
  - How are the students' thinking and reasoning the same and how are they different between the two Leftover Pizza lessons?

Discuss your responses with your collaborative team before moving on to the activity 2.1 discussion.

Leftover Pizza Lesson Version 1:
www.SolutionTree.com/Dividing_Fractions_in_Context

A teacher reads the Leftover Pizza task to the students, then asks, "So what are we trying to find out?" A student responds, "How many servings can he make if he uses up all the pizza." The following dialogue ensues, with the teacher calling on different students in the class to answer her questions.

Teacher: How much pizza is in each serving?

Student: There is $2/3$ of a pizza in each serving.

Teacher: So the problem is asking you to determine how many $2/3$ are in $4\,5/6$. Is this multiplication or division?

Student: Division.

Teacher: Correct, so your task is to compute $4\,5/6$ divided by $2/3$. What is the first thing you need to do?

Student: Change $4\,5/6$ to an improper fraction.

Teacher: Improper fractions are also called "fractions greater than one." How could we rewrite $4\,5/6$ as a fraction greater than one?

Student: We could write $29/6$.

Teacher: That's right. Now we have $29/6$ divided by $2/3$. How do we divide fractions class?

Class (*enthusiastic choral response*): Keep-change-flip!

Teacher: You got it! After we keep-change-flip, we have $29/6$ times $3/2$ (*teacher writes $29/6 \div 2/3 = 29/6 \times 3/2$ on the board*). Write this on your whiteboards, then multiply the fractions to get your answer. After you have your answer, compare with your partner.

> The students work independently and then share their answers with their partners. The teacher calls the students' attention back to the front of the room.
>
> **Teacher:** What did you get?
>
> **Student:** I got 7¼.
>
> **Teacher:** 7¼ what?
>
> **Student:** 7¼ servings.
>
> **Teacher:** That's right. Don't forget to go back to the problem so you use the correct unit in your answer. Nice work!

Figure 2.1: Leftover Pizza lesson version 2.

Discuss

How do your responses compare with those in your collaborative team? What themes emerged during your discussion? In this section, we present ideas for you to consider.

Rate the Leftover Pizza task using the Potential of the Task rubric.

The Leftover Pizza task would rate a level 3 on the Potential of the Task rubric. As discussed in activity 1.1, the task provides a context and encourages a model for students to make sense of dividing fractions. The task does not include an explicit prompt to explain.

What do you notice about the Leftover Pizza lesson version 1? How would you describe the students' thinking and reasoning during the lesson?

The students grapple with a challenging problem and have materials such as pattern blocks and whiteboards that support their engagement with the task. Students communicate mathematically with peers and have opportunities to make sense of each other's mathematical thinking. The teacher consistently requests explanation and meaning when students give their solution methods and debate 7⅙ versus 7¼. As noted in chapter 1, this task is designed to elicit a dilemma and common misconception in interpreting fraction division—the remaining piece is ⅙ of a pizza, but ¼ of a serving.

Several students explain their thinking during the lesson. The first student explains how he got 7⅙ by making 7 servings of pizza with ⅙ of the pizza left over, so the solution is 7 and ⅙ (3:12). A second student disagrees but cannot explain the difference between the serving and the leftover slice (5:18). The teacher calls on a student whom she had heard talk about servings, and the student explains that ⅙ of the pizza is ¼ of a serving (5:33). The teacher asks if the second student can make sense of that explanation, and he explains that because four triangles make up a serving, one triangle would be ¼ of a serving (5:52). The teacher then goes back to the original student (who expressed the common but incorrect way of interpreting the remaining piece of pizza), and that student is now able to say that the ⅙ piece of pizza is ¼ of a serving (6:00).

What do you notice about the Leftover Pizza lesson version 2? How would you describe the students' thinking and reasoning in the narrative?

Students engage in using a procedure for dividing fractions that the teacher specifically calls for. There is little ambiguity about what they need to do in solving the problem and how to do it. Students do not make connections to how the problem relates to division of fractions, creation of equal groups, or the meaning underlying the procedure they are using. Implementation focuses on producing a correct answer rather than developing mathematical understanding—in this case, using an algorithm for dividing fractions. While students get access to the problem by receiving a procedure to use for the solution, there is no evidence of student understanding.

How are the students' thinking and reasoning the same and how are they different between the two versions of the Leftover Pizza lesson?

In both lessons, students arrive at the correct answer of 7¼. In version 1, students use the context and create models to make sense of dividing fractions, particularly of the remaining part of a serving. While the task does not include an explicit prompt to explain, the teacher verbally prompts students to explain. Hence, the teacher enhances students' opportunities for thinking and reasoning, beyond the opportunities provided by the original task, by asking for explanations.

In version 2, the teacher directs students toward a set procedure for dividing fractions. The teacher sets up the computation for students at the board and summarizes exactly what students should do next: "multiply the fractions to get your answer." The teacher does not encourage students to use the context and models to make sense of the division of fractions. Instead, the teacher limits students' mathematical activity to performing a computation, thereby reducing the opportunities for thinking and reasoning provided by the task. This process offers the opportunity for all students to engage with the lesson and even reach the correct answer for this single problem; however, there is no evidence that students understand the procedure for dividing fractions or that they will be able to solve a similar problem without explicit prompts from the teacher. There is also no indication that they will know when to apply the procedure given a fraction operation problem in context, thereby denying them access to future problems. This is in contrast to the access the teacher provides in version 1 of the lesson, in which students could use manipulatives and make sense of the context, and use that understanding to solve this problem and have access to future problems.

As demonstrated by the two versions of the Leftover Pizza lesson, the same task can elicit different levels and types of thinking depending on how the task is implemented. Tasks with the potential to elicit thinking and reasoning (levels 3 and 4) can be implemented in ways that support students' learning of mathematics with understanding (as in the Leftover Pizza lesson version 1) or in ways that support procedural or rote learning (as in the Leftover Pizza lesson version 2).

In chapter 1, you used the Potential of the Task rubric to rate the opportunities for thinking and reasoning embedded in an instructional task. Next, you will explore the IQA Implementation of the Task rubric, which will enable you to rate the level and type of thinking students actually exhibit throughout a mathematics lesson.

The IQA Implementation of the Task Rubric

The IQA Implementation of the Task rubric (figure 2.2) is intended to rate students' overall engagement in thinking and reasoning throughout an entire lesson. The Implementation of the Task rubric rating indicates whether the level of cognitive demand of the instructional task, as determined by the Potential of the Task rubric, was maintained, increased, or decreased throughout the lesson. In other words, once you rate the level of *potential* opportunities for thinking and reasoning that an instructional task can provide, you then assess the level of *actual* thinking and reasoning that takes place during a lesson. For this reason, the wording of the Implementation of the Task rubric parallels the wording of the Potential of the Task rubric, with the verbs changed to past tense.

Implementation of the task is rated based on lesson observations, video recordings of instruction, or sets of students' work. The Implementation of the Task rubric is a holistic rating of what most of the observed students are doing most of the time. (In later chapters, we introduce additional rubrics that more closely analyze the students' contributions to the lesson, the teacher's questioning, and the quality of mathematical discussions.)

As you read the Implementation of the Task rubric, consider how you would rate each version of the Leftover Pizza lesson.

	IQA Implementation of the Task Rubric
4	Students engaged in exploring and understanding the nature of mathematical concepts, procedures, or relationships. There is *explicit evidence* of students' reasoning and understanding. For example, students may have: • Solved a genuine, challenging problem for which students' reasoning is evident in their work on the task • Developed an explanation for why formulas or procedures work • Identified patterns and formed and justified generalizations based on these patterns • Made conjectures and supported conclusions with mathematical evidence • Made explicit connections between representations, strategies, or mathematical concepts and procedures • Followed a prescribed procedure in order to explain or illustrate a mathematical concept, process, or relationship
3	Students engaged in complex thinking or in creating meaning for mathematical concepts, procedures, or relationships. However, the implementation does not warrant a level 4 because there were *no explicit explanations or written work* to indicate students' reasoning and understanding. Students may have: • Engaged in problem solving, but for a task that required minimal cognitive challenge (for example, the problem was easy to solve), or students' reasoning is not evident in their work on the task • Explored why formulas or procedures work but did not provide explanations • Identified patterns but did not form or justify generalizations • Made conjectures but did not provide mathematical evidence or explanations to support conclusions • Used multiple strategies or representations but connections between different strategies or representations were not explicitly evident • Followed a prescribed procedure to make sense of a mathematical concept, process, or relationship, but did not explain or illustrate the underlying mathematical ideas or relationships

Source: Adapted from Boston, 2017.

Figure 2.2: IQA Implementation of the Task rubric.   continued →

2	Students engaged in using a procedure that either was specifically called for or its use was evident based on prior instruction, experience, or placement of the task. • There was little ambiguity about what needed to be done and how to do it. • Students did not make connections to the concepts or meaning underlying the procedure being used. • The focus of the implementation appeared to be on producing correct answers rather than developing mathematical understanding (for example, applying a specific problem-solving strategy or practicing a computational algorithm).
1	Students engaged in memorizing; note taking; or reproducing facts, rules, formulas, or definitions. Students did not make connections to the concepts or meanings that underlie the facts, rules, formulas, or definitions being memorized or reproduced.
0	Students did not engage in mathematical activity.

Recall the Leftover Pizza task rated a level 3 on the Potential of the Task rubric. In version 1, we would rate the lesson as a level 4 on the Implementation of the Task rubric, because students provide explanations during the whole-group discussion in response to prompts from the teacher. We would consider version 1 of the Leftover Pizza lesson as increasing the cognitive demand from a level 3 on the Potential of the Task rubric to a level 4 on the Implementation of the Task rubric. We would consider version 2 of the Leftover Pizza lesson as decreasing the cognitive demand from a level 3 on the Potential of the Task rubric to a level 2 on the Implementation of the Task rubric, because the focus of students' work shifts to performing procedures and obtaining correct answers.

In the following application activities, you will have the opportunity to use the Implementation of the Task rubric on students' work.

Application Activities

The following activities will help you become familiar with the IQA Implementation of the Task rubric as you analyze mathematics lessons and students' mathematical work.

Activity 2.2: Using the Implementation of the Task Rubric

It is valuable to engage with tasks as learners to make sense of what those tasks have to offer students. Be sure to devote attention to this experience. Explore the tasks that are new to you on your own before engaging in the activity.

Engage

In this activity, you will consider the benchmark samples of students' work in figure 2.3 and give each set of students' work, representing one lesson, one rating for the Implementation of the Task rubric based on the evidence of thinking and reasoning in the majority of examples of students' work.

- Use the Potential of the Task and Implementation of the Task rubrics to justify the rating of each task and student work sample in figure 2.3. The tasks are from varied lessons and grade levels and are provided to illustrate student work representing different levels of implementation. (Note that when using the Implementation of the Task rubric to rate student work, you would typically look over a set of student work, rather than one exemplar as provided here.)

- Discuss your ideas with your collaborative team.

Sample Student Work	Justification
Level 4: Revised Leftover Pizza Task (Grade 6) Douglas ordered 5 small pizzas during the great pizza sale. He ate ⅙ of one pizza and wants to freeze the remaining 4⅚ pizzas. Douglas decides to freeze the remaining pizza in serving-size bags. A serving of pizza is ⅔ of a pizza. How many servings can he make if he uses up all the pizza? Explain your reasoning. 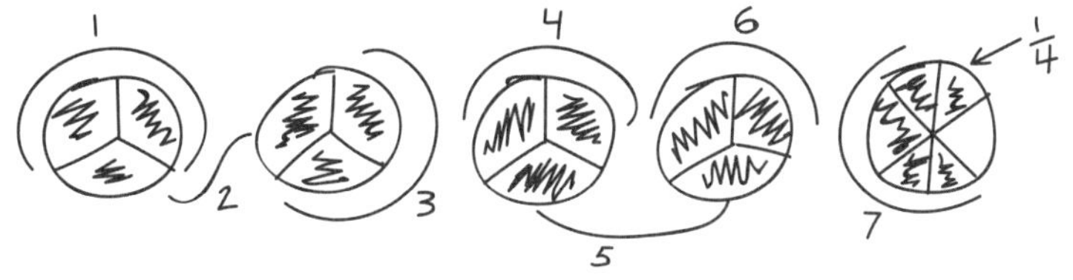 *[student diagram of 5 circles divided into thirds with servings labeled 1–7 and ¼]* I know that there are 7 full servings of pizza in 4⅚ pizzas. There is ⅙ of a pizza that doesn't fit in a full serving. Because it takes 4/6 of a pizza to make a serving, then the ⅙ of a pizza is ¼ of a serving. Douglas can make 7¼ servings from the leftover pizza.	
Level 3: Fraction Operation Model Task (Grade 4) Write a number sentence that can be represented by this model. 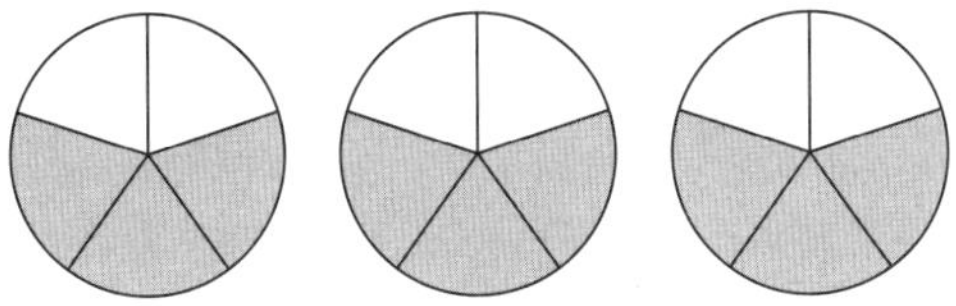 *[three circles each divided into 5 parts with 3 parts shaded]* $3 \times \frac{3}{5} = 1\frac{4}{5}$	

Level 2: Simplifying Expressions Task (Algebra)
Simplify the following expressions. Remember to combine like terms if possible.
 	
1. $-8x^2 - 5 + 2x^2 + 12 - 6x^2$
   $-12x^2 + 7$
 	
2. $2abc - 9a^2bc - 8abc - 12ab - 16$
   $-6abc - 9a^2bc - 12ab - 16$
 	
3. $-6(2x + 3) + 2(-7 + 3x)$
   $-12x - 18 - 14 + 6x$
   $= -6x - 32$	

Source: Questions adapted from Nolan, Dixon, Roy, & Andreasen, 2016, p. 38.

Figure 2.3: Benchmark samples of implementation.

continued →

Sample Student Work	Justification

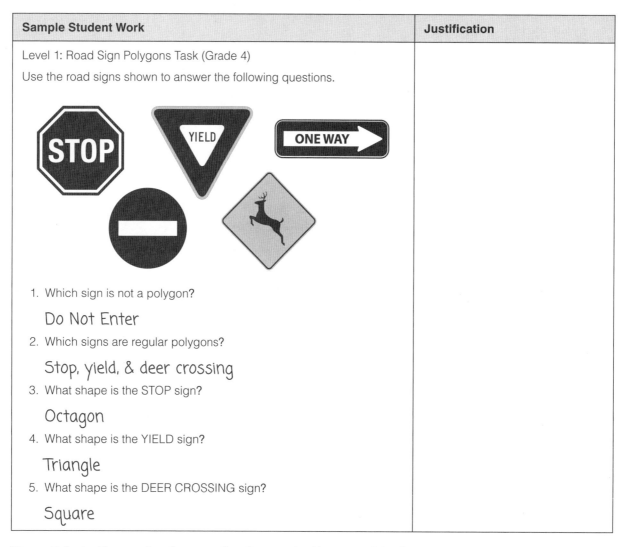
Level 1: Road Sign Polygons Task (Grade 4)
Use the road signs shown to answer the following questions.
 	
1. Which sign is not a polygon?
   Do Not Enter
2. Which signs are regular polygons?
   Stop, yield, & deer crossing
3. What shape is the STOP sign?
   Octagon
4. What shape is the YIELD sign?
   Triangle
5. What shape is the DEER CROSSING sign?
   Square	

*Visit **go.SolutionTree.com/mathematics** for a free reproducible version of this figure.*

Discuss

How do your responses compare with those in your collaborative team? What themes emerged during your discussion? In this section, we present ideas for you to consider.

Use the Potential of the Task and Implementation of the Task rubrics to justify the rating of each task and student work sample in figure 2.3.

Here we provide our justifications for the rating of each task and student work.

- **Level 4:** In the student work for the revised Leftover Pizza task, the student's thinking and reasoning is evident. The student provides a solution for the task—both visually and in writing—and provides a justification for the size of the partial serving. The student does not simply indicate that the answer to the problem is 7¼ servings.

  For a lesson to be rated a level 4 on the Implementation of the Task rubric, students must provide explanations or other explicit evidence of their thinking and reasoning. The nature of explanations will differ according to student age and the mathematical topics. A picture might serve as a complete explanation at grades K through 1, in which students are in the process of developing language and writing skills. An annotated picture or diagram might also provide a thorough explanation of students' thinking and reasoning for a particular mathematical task or topic across grade levels. Sometimes it is difficult to rate tasks and explanations that engage students in making sense of procedures. For example, as students are developing their own strategies and understanding of multidigit subtraction, explanations of their thinking may sound very procedural, because they are thinking and reasoning about mathematical procedures.

- **Level 3:** In the student work for the Fraction Operation Model task, the student provides a number sentence that represents the given model. The student recognizes the appropriate fraction and operation exhibited by the model. However, the student does not explain the connection between the model and the number sentence. While there is an indication that the student creates meaning for a mathematical process, there is no explicit explanation or written work to indicate the student's reasoning or understanding. Implementation would increase to a level 4 if the student added, "Because there are three groups of ⅗, that would be the same as ⁹⁄₅, or 1⅘," explaining the meaning of multiplication. Note that there is no set procedure for solving the task, and the goal is not to perform a computation and produce an answer.

    For an Implementation of the Task rubric rating at a level 3, students exhibit an underlying understanding of mathematical concepts, procedures, or relationships. Often, solving the task requires mathematical connections or sense making, and students demonstrate the ability to solve the task but do not provide justifications or explanations.

- **Level 2:** In the student work for the Simplifying Expressions task, the student uses a variety of procedures for simplifying expressions, including the distributive property and combining like terms. The student solves several problems of the same type, with slight variations. Some steps of the procedures are evident, and some steps may be performed mentally. Even so, the student applies procedures to obtain answers, and the Implementation of the Task rubric rates a level 2.

    For a level 2 on the Implementation of the Task rubric, students' mathematical activity is characterized by procedures or computations. Students often complete several problems requiring the same type of procedure, perhaps with small variations. Often, the teacher has demonstrated or modeled this procedure for students prior to their opportunity to do the work. This includes students' completion of word problems or real-life applications in which the procedure is indicated (or evident) prior to students' work, and each word problem requires the same operation or procedure to solve. If students can solve a word problem without having to actually read the words in the problem (or make sense of the contextual situation), the problem is merely procedural.

- **Level 1:** In the student work for the Road Sign Polygons task, the student provides single-word responses that indicate recognition of shapes or properties of shapes. Implementation rates a level 1 because the student engages in reproducing facts or definitions. In general, in lessons with a rating of level 1 on the Implementation of the Task rubric, students (or students' work) provide evidence of recalled or reproduced information. There is no mathematical procedure involved. Lessons where students' main mathematical activity is note taking are also rated a level 1 on the Implementation of the Task rubric because students are reproducing (writing down) someone else's mathematical ideas, procedures, examples, or definitions. Directing students to copy the steps of a mathematical procedure is very different from having students complete the procedure themselves.

Figure 2.3 (page 35) is helpful for providing exemplars of tasks and student work at each of the four levels. As you saw with the Leftover Pizza task and lessons, however, tasks are frequently implemented at different levels than their potential—both higher and lower. In the next activity, you will have the chance to use the Implementation of the Task rubric to consider whether the cognitive demands of mathematical tasks are maintained, increased, or decreased during implementation by comparing the rating on the Potential of the Task rubric to the rating on the Implementation of the Task rubric.

Activity 2.3: Identifying Evidence of Students' Thinking and Reasoning in Students' Work

Student work can be helpful in reflecting on whether the potential of the task was maintained or changed. In this activity, we ask you to look through a collection of student work to examine the level of implementation.

Engage

Figure 2.4 identifies three tasks from the Benchmark Tasks grid in chapter 1 (figure 1.2, page 13). Figures 2.5 (page 40), 2.6 (page 41), and 2.7 (page 42) provide sets of student work for each of the three tasks.

- Use the Implementation of the Task rubric to rate the level of students' engagement with each task. Look across the set of students' work for each task, and give the entire set one rating based on the majority of the evidence.

- Compare the Implementation of the Task rating to the rating for Potential of the Task (rating for the Potential of the Task provided in figure 2.4). Determine if the cognitive demands of the task were maintained, increased, or decreased and provide reasons.

Tasks (from figure 1.2)	Potential of the Task	Student Work
Write a word problem for 26 divided by 4 that results in an answer of 7. Do not use the words *around*, *estimate*, or *about*.	3	See figure 2.5: 26 Divided by 4
A trapezoid is shown below. Using any combination of rectangles, parallelograms, and triangles, determine a formula for the area of this trapezoid. Justify why your formula works. [Trapezoid with top base b_1, bottom base b_2, and height h]	4	See figure 2.6: Area of a Trapezoid
Copy in your notes the rules for determining the sign of the sum of two integers: a. Positive + Positive → Positive b. Negative + Negative → Negative c. Positive + Negative or Negative + Positive → Sign of the integer with the larger absolute value	1	See figure 2.7: Rules for Integers

Source: Questions adapted from Dixon, Nolan, Adams, Tobias, & Barmoha, 2016, p. 60; Nolan, Dixon, Roy, & Andreasen, 2016, p. 105.

Figure 2.4: Tasks for activity 2.3.

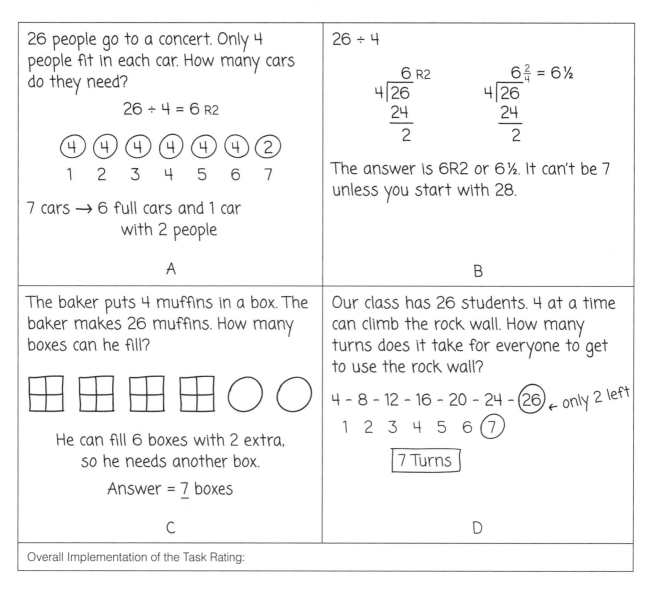

Figure 2.5: Student work samples for 26 Divided by 4 task.

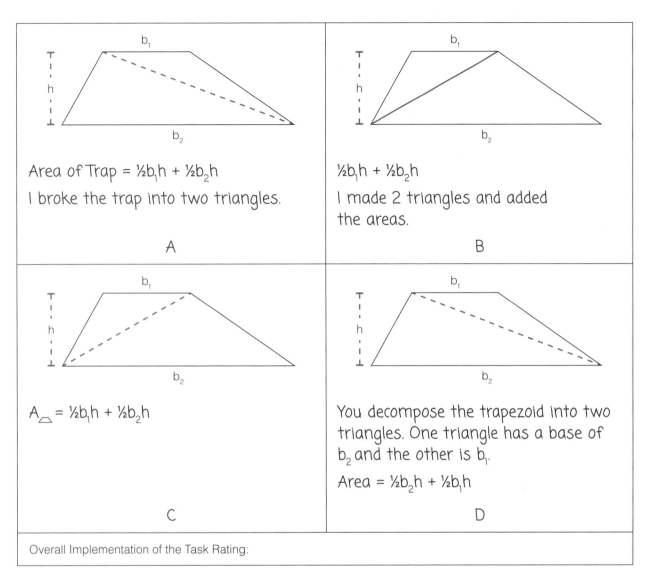

Figure 2.6: Student work samples for Area of a Trapezoid task.

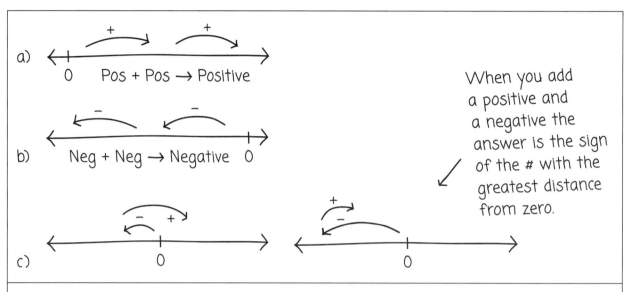

Figure 2.7: Student work samples for Rules for Integers task.

Discuss

How do your responses compare with those in your collaborative team? What themes emerged during your discussion? In this section, we present ideas for you to consider.

Use the Implementation of the Task rubric to rate the level of students' engagement with each task and provide reasons for your rating. Compare the Implementation of the Task rating to the rating for Potential of the Task. Determine if the cognitive demands of the task were maintained, increased, or decreased.

For each set of student work, we discuss the level of implementation of the task.

- **26 Divided by 4 student work (figure 2.5, page 40):** Implementation of the Task rates a level 4. Students create situations for 26 ÷ 4 that result in an answer of 7, and they provide explanations for why the result is 7. This is true even though not all of the student work samples have the correct answer (for example, student B). However, the task rated a level 3 on the Potential of the Task rubric, as the task did not explicitly request explanations. Hence, students' level of thinking and reasoning increased during implementation by including explanations. This increase from a level 3 on the Potential of the Task rubric to a level 4 on the Implementation of the Task rubric is apparent in classrooms where teachers hold students accountable for thinking and reasoning even when such prompts are absent from the original tasks.

- **Area of a Trapezoid student work (figure 2.6, page 41):** Implementation of the Task rates a level 2. All student work looks similar, indicating the application of a prescribed or demonstrated procedure. The trapezoids are all divided into two triangles, and the computations are almost exactly the same. Student work does not contain justifications for why the formula works, even though explicitly called for by the task. Often, when tasks at a level 3 or 4 for Potential of the Task are solved during whole-group instruction or through direct modeling, student work will look the same or very similar. In these situations, students do not have the opportunity to engage with the demands of the task and their engagement is often reduced to performing computations and arriving at an answer. Alternatively, this pattern of student work could be the result of students having performed a particular problem several times—thus relegating it to a more procedural problem. In either situation, the cognitive demands decline during implementation.

  A decline from a Potential of the Task rating of level 3 or 4 to an Implementation of the Task rating of level 2 is the most common pattern observed in research (Boston & Wilhelm, 2015). As in version 2 of the Leftover Pizza lesson, well-intentioned teachers might decide to walk students through the difficult parts of a task. While this over-scaffolding may lead to a correct solution to the specific problem at hand, it limits students' engagement with the problem-solving or sense-making aspects of the task and does not promote their understanding of larger mathematical ideas.

- **Rules for Integers student work (figure 2.7):** Implementation of the Task is rated a level 4. While the original task suggested the copying of notes (Potential of the Task rating of level 1), student work provides evidence of thinking and reasoning as students provide unique models and explanations for each rule. Increases from a Potential of the Task rating of level 1 or 2 to an Implementation of the Task rating of level 3 or 4 are the least common patterns observed in classroom research (Boston & Wilhelm, 2015; Henningsen & Stein, 1997). However, well-timed questions from a teacher or student ("How do we know this procedure will always work?"), a move to understand a procedure ("Which method works best for this type of problem? Why?"), the requirement of a model or explanation, or a request to develop a procedure, generalization, or proof ("What do you notice about the sum when both addends are odd? Why? Will this always be the case?") can supplement the original focus on procedures and memorized knowledge by introducing an opportunity for thinking and reasoning.

In summary, collections of students' work provide evidence about whether the cognitive demands of tasks are maintained, increased, or decreased as a task is implemented during instruction. Rating the implementation of the task based on student work provides an indication of students' level of engagement with the task and can also provide information about the teaching, instruction, or feedback that occurs during the lesson.

When rating a set of students' work, consider whether students solve the task in more than one way or by using more than one representation (across the set of samples or within individual samples), provide explanations, or engage with the cognitively challenging parts of the task (or leave those parts blank or incomplete). Figure 2.8 lists characteristics of student work when the cognitively challenging demands of the task levels have been maintained, increased, or decreased during implementation.

Maintain the demand.
- Student work provides evidence that students have met the cognitively challenging aspects of the task.
- **Level 4:** Students provide (at least) adequate explanations.
- **Level 3:** Students provide evidence of engagement in thinking and reasoning, but students' responses require teachers to make inferences about what exactly students understand because no explanations are provided.

Increase the demand.
- **Level 3 to 4:** Students provide explanations, even when the original task did not explicitly prompt for explanations.
- **Level 2 to 3:** Students provide multiple strategies or representations, or there are unique strategies across the set of students' work, even when the original task did not prompt for multiple strategies or representations, or appeared to have a set strategy or procedure.
- In general, increases from Potential to Implementation:
  - Might be evidence of established norms in the classroom
  - Might reflect ideas generated during the lesson (discussions or group work)
  - Might be in response to teacher questioning or feedback during the lesson

Decrease the demand.
- Student work all looks the same or follows a similar template; explanations are similarly worded.
- Students do not provide explanations, or students skip the challenging parts of the task.
- In general, decreases from Potential to Implementation:
  - Might indicate the teacher's feedback, modeling, or questions were too directive
  - Might indicate one student took over during group work

Figure 2.8: Characteristics of student work when cognitive demand is maintained, increased, or decreased.

*Visit **go.SolutionTree.com/mathematics** for a reproducible version of this figure.*

Activity 2.4: Revisiting the Chapter 1 Transition Activity

Student work can be helpful in reflecting on whether the potential of the task was maintained or changed. In this activity, we ask you to look through a collection of student work from your own classroom to examine the level of implementation.

Engage

In the chapter 1 transition activity (page 27), you collected copies of three instructional tasks and identified features that helped the tasks rate a level 3 or 4. You also collected sets of student work from each task (at least four samples each) that included a variety of strategies, thinking, and reasoning.

- Use the Implementation of the Task rubric to rate the sets of students' work.
- Was the Potential of the Task level maintained, increased, or decreased? Why?

Discuss

Using the student work you collected, did you and your collaborative team identify some of the same characteristics identified in figure 2.8? Did you find it difficult or uncomfortable to discuss instances in which cognitive demands declined? We acknowledge that reflecting on instruction (your own or someone else's) might identify some areas of improvement that are difficult to consider or discuss. In our work with teachers, analyzing student work using the IQA rubrics has served as a good starting point for teachers to share their classroom practices. For difficult conversations or self-reflections, we offer the following encouragement.

- In research on the use of cognitively challenging tasks, the greatest impact on students' learning occurred in classrooms where teachers consistently maintained the demand of the task during implementation (Cai et al., 2013; Grouws et al., 2013). However, students in classrooms where teachers used cognitively challenging tasks but the cognitive demands consistently declined during implementation still exhibited moderate gains in student achievement, statistically higher than students in classrooms where teachers did not use cognitively challenging tasks (Boaler & Staples, 2008; Stein & Lane, 1996).

- In a professional development workshop focused on implementing cognitively challenging tasks, teachers who were the most open and vocal about their practice and their own areas of needed improvement made significant changes in the levels of their tasks' potential and implementation throughout the experience (Boston & Smith, 2009) and even two years later (Boston & Smith, 2011).

- Teachers who used the IQA rubrics as a tool to analyze and reflect on their practice in a professional development initiative expressed that having a shared language and set of ideas was helpful for considering their own growth and providing feedback to colleagues (Boston & Candela, 2018; Candela, 2016).

In the next activity, you will explore how to rate Implementation of the Task based on observations or videos.

Considerations for Implementation Based on Classroom Observations or Videos

The IQA was created as a tool that teachers could use reliably during classroom observations. When rating Implementation of the Task based on a lesson observation or video, consider the entire lesson from the point students begin to work on a task (following any directions by the teacher) and include any student work time and discussion (small-group and whole-group) until the end of the lesson. Look for evidence of thinking and reasoning in students' written work, small-group work and discussions, and any whole-group discussions. Implementation of the Task is based on a holistic rating of what most of the students are doing most of the time, or what level seems to characterize the lesson, as students may be engaged in different levels and types of thinking throughout the lesson.

When applying the Implementation of the Task rubric during classroom observations or while watching a video, use the Implementation Observation Tool (figure 2.9) to keep notes and identify instances of when you notice an instructional move in the classroom. Sections A and B allow you to identify where the lesson does or does not provide opportunities for students to engage with high-level cognitive demand. In section C, note instances when the student discussion provides opportunities for students to engage with the high-level demands of the task. You may also find it helpful to take running notes on instances when teachers and students contribute to the maintenance or decline of the cognitive demand of the task.

There is no rigid formula for moving from the Implementation Observation Tool to the Implementation of the Task rubric; however, a rating of level 4 should be characterized by observance of multiple items in section A, along with strong evidence of at least one item in section C on student discussion, as explicit evidence of connections and reasoning by students. A rating of level 3 will also be characterized by items in section A, but would not have strong evidence of items in section C (indicating a lack of explicit evidence of students' opportunities to make connections, justifications, and explanations). A rating of level 2 will be characterized by items in section B, which indicate diminished opportunities for students to engage in a task at a high level.

In the next activity, and throughout the rest of the book, you will watch short video clips of mathematics lessons. Note that each video clip, approximately four to eight minutes in length, illustrates a specific point about a rubric or a topic for discussion. Because you are viewing a small excerpt of a lesson, the lesson may feel incomplete in the sense that you may want to observe other parts of the lesson or have the teacher ask additional questions, discuss additional mathematical points, or hear additional students' responses. For each video, consider the teacher's actions and evidence of students' thinking and reasoning that you observe in the brief video clip, rather than the actions, thinking, and reasoning that are not included.

Implementation of the Task

IQA Implementation Observation Tool	
A ↑ The lesson **provides** opportunities for students to engage with high-level cognitive demand.	**B** ↓ The lesson **does not provide** opportunities for students to engage with high-level cognitive demand.

• Students:
  ○ Engage with the task in ways that address the teacher's goals for high-level thinking and reasoning
  ○ Communicate mathematically with peers
  ○ Have appropriate prior knowledge to engage with the task
  ○ Have opportunities to serve as mathematical authorities in the classroom
  ○ Have access to resources that support their engagement with the task
• The teacher:
  ○ Supports students to engage with the high-level demands of the task while maintaining the challenge of the task
  ○ Provides sufficient time to grapple with the demanding aspects of the task and to expand thinking and reasoning
  ○ Holds students accountable for high-level products and processes
  ○ Provides consistent requests for explanation and meaning
  ○ Provides students with sufficient modeling of high-level performance on the task
  ○ Provides encouragement for students to make conceptual connections
	• The task:
	  ○ Expectations are not clear enough to promote students' engagement with the high-level demands of the task
	  ○ Is not rigorous enough to support or sustain student engagement in high-level thinking
	  ○ Is too complex to sustain student engagement in high-level thinking (Students do not have the prior knowledge necessary to engage with the task at a high level.)
	• The teacher:
	  ○ Allows classroom management problems to interfere with students' opportunities to engage in high-level thinking
	  ○ Provides a set procedure for solving the task
	  ○ Shifts the focus to procedural aspects of the task or on correctness of the answer rather than on meaning and understanding
	  ○ Gives feedback, modeling, or examples that are too directive or do not leave any complex thinking for the student
	  ○ Does not press students or hold them accountable for high-level products and processes or for explanations and meaning
	  ○ Does not give students enough time to deeply engage with the task or to complete the task to the extent that is expected
	  ○ Does not provide students access to resources necessary to engage with the task at a high level

C The **discussion** provides opportunities for students to engage with the high-level demands of the task.	

Students:

- Use multiple strategies and make explicit connections or comparisons between these strategies, or explain why they chose one strategy over another
- Use or discuss multiple representations and make connections between different representations or between the representation and their strategy, underlying mathematical ideas, or the context of the problem
- Identify patterns or make conjectures, predictions, or estimates that are well grounded in underlying mathematical concepts or evidence
- Generate evidence to test their conjectures and use this evidence to generalize mathematical relationships, properties, formulas, or procedures
- Determine the validity of answers, strategies, or ideas rather than waiting for the teacher to do so

Source: Adapted from Boston, 2017.

Figure 2.9: Implementation Observation Tool.

Activity 2.5: Rating Implementation of the Task—Father and Son Race Lesson Version 1

In this activity, you will engage with a mathematics task, consider the cognitive demands of the task, and explore the implementation of the task.

Engage

It is valuable to engage with tasks as learners to make sense of what those tasks have to offer students. Be sure to devote attention to this experience. Explore the task on your own before engaging in the activity.

- Rate the Father and Son Race task (see figure 2.10) using the Potential of the Task rubric. Be sure to include your rationale—what characteristics or features influenced your rating of the task? Discuss your rating with your collaborative team before moving on.

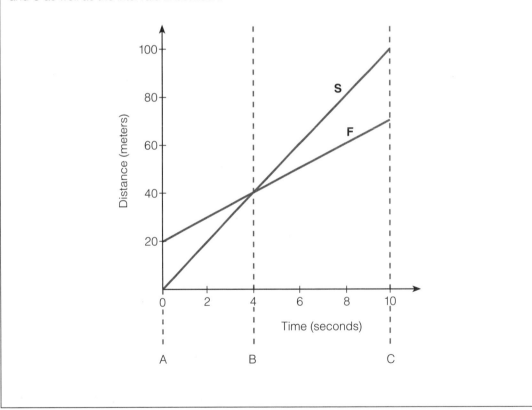

The Father (F) and Son (S) 100-Meter Race (Algebra)

The figure below shows the relationship of distance and time for a father (F) and son (S) during a 100-meter race. Write a story that matches the graph. Be sure to include what is occurring at A, B, and C as well as the intervals in between.

Source: Nolan, Dixon, Safi, & Haciomeroglu, 2016, p. 16.

Figure 2.10: The Father and Son Race task (algebra).

*Visit **go.SolutionTree.com/mathematics** for a free reproducible version of this figure.*

- Next, watch the Father and Son Race lesson version 1. Use the Implementation of the Task rubric to rate the level of thinking and reasoning students exhibited during the lesson.
    - Did students appear to be making sense of linear relationships?
    - Did you hear evidence of students explaining their thinking about linear relationships or justifying their ideas about the context of the race?
    - In what ways did the task and teacher support access to the task?
    - How does the Implementation of the Task rubric rating compare to the Potential of the Task rubric rating? What aspects of the lesson made the Implementation of the Task rating the same as or different from the Potential of the Task rating?

Father and Son Race Lesson Version 1:
www.SolutionTree.com/Interpreting_Graphs_to_Create_Context

Discuss

How do your responses compare with those in your collaborative team? What themes emerged during your discussion? In this section, we present ideas for you to consider.

Rate the Father and Son Race task using the Potential of the Task rubric.

The Father and Son Race task rates a level 3 for Potential of the Task. The task provides the opportunity for students to make sense of many aspects of linear relationships (such as y-intercept, point of intersection, and rate of change) and to describe what these ideas mean in context. While writing a story will provide implicit evidence of students' understanding (for example, we could assume students understand ideas about linear relationships if they represent these ideas correctly within the context), the task does not explicitly prompt students to explain their thinking or decision making in creating the story; therefore, it does not rate a level 4.

Use the Implementation of the Task rubric to rate the level of thinking and reasoning students exhibited during the lesson.

Students show thinking and reasoning by explaining the meaning of the y-intercept, point of intersection, and rate of change. Implementation is also impacted by the actions of the teacher. The teacher in the lesson provides students time to make sense of this task. She circulates around the room to determine what students are doing well and where they are making errors (though note that she does not always share with students what she notices). She questions students at the small-group level and then pulls the class back for a whole-class discussion regarding the task. During the whole-group discussion, she continues to provide opportunities for students to explain their thinking (for example, by asking, "What on the graph told you that? What is happening at B? What is happening with the speed?"). When a misconception surfaces (4:40), the teacher asks the class whether they agree or disagree and why. She returns to the student later in the lesson and indicates that the student's previous idea is a common misconception in interpreting linear relationships.

The Implementation of the Task in version 1 would be rated a level 4 because students engage with the task and communicate mathematically with their peers about the situation in the graph. The students have appropriate prior knowledge of graphs to engage in this task and have opportunities to serve as mathematical authorities when talking to peers in small groups and in the whole-class discussion. The teacher supports engagement with high-level aspects of the task by asking questions regarding students' interpretations of the graphs. The teacher holds students accountable for their statements and follows up with questions so that students make connections between their statements and the graphs. The students rather than the teacher determine the validity of the answers and are able to realize that it is the distance that is changing in the graph, not the speed of the father and son.

Activity 2.6: Rating Implementation of the Task—Father and Son Race Lesson Version 2

Imagine if the teacher in the Father and Son Race lesson version 1 had implemented the task differently.

Engage

Read version 2 of the Father and Son Race lesson in figure 2.11 and consider the following questions.

- How do you rate version 2 of the lesson using the Implementation of the Task rubric? What is the level of thinking and reasoning exhibited by students during the lesson?
- How does the scenario in version 2 provide different opportunities for students' thinking and reasoning than in version 1?
- How does the scenario in version 2 affect the level of access each student has to the task compared to the scenario in version 1?

Teacher: Your goal is to write a story to match this graph. Notice that at A, the son is at zero seconds and zero meters while the father is given a head start and is at 20 meters at the start of the race. What do we call the point (0, 20), Delta?

Delta: The *y*-intercept.

Teacher: That's right. What do we call the point at B, Calvin?

Calvin: The point of intersection.

Teacher: That's right. The point of intersection tells us when the father and son are at the same time and the same distance. Now I have a more challenging question for you. Notice that the lines on the graph for the father and for the son are both straight lines. What does that indicate about the speeds of the father and the son? Before you answer, think about what a straight line tells us about the rate of change. Blaize?

Blaize: A straight line tells us that there is a constant rate of change. So I guess that means the father and the son are running at a constant speed. But it looks like the son starts to run faster after point B.

Teacher: That's right, it does look like that, but it is not correct. The speeds never change. Now that you have made sense of the graph, please write your stories on your own and be prepared to turn in your paper by the time the bell rings.

Figure 2.11: Father and Son Race lesson version 2.

Discuss

How do your responses compare with those in your collaborative team? What themes emerged during your discussion? In this section, we present ideas for you to consider.

How do you rate version 2 of the lesson using the Implementation of the Task rubric? What is the level of thinking and reasoning exhibited by students during the lesson?

In version 2 of the Father and Son Race lesson, the teacher, rather than the students, makes the mathematical connections. In the launch of the task, the teacher tells students the father has a 20-meter head start and brings students' attention to the initial difference between the starting point of the father and the son. The teacher asks very specific questions about the graph, leaving little ambiguity about what needs to be done and how to do it. There is no evidence of students making connections between the points of the graph and what is happening in the race. Instead, the teacher makes those connections and explains to the students what the point of intersection means as well as how rate of change relates to the graph. When Blaize starts to suggest the son runs faster than the father, the teacher simply tells students that speed never changes.

Implementation in the Father and Son Race lesson version 2 is focused on producing correct answers rather than developing mathematical understanding. The teacher is looking for students to say what the point of intersection *is*, rather than what that point of intersection *means*. For these reasons, the Implementation of the Task in version 2 is rated at a level 2.

How does the scenario in version 2 provide different opportunities for students' thinking and reasoning than version 1?

In version 2 of the Father and Son Race lesson, students do not receive opportunities to think and reason about the task. The teacher does the thinking and reasoning about the main features of the task and merely wants to know if students can correctly identify these features. The focus is on procedural aspects of giving the definitions of intersection and rate of change, rather than on making connections to what the intersection and rate of change mean and how they relate to this particular context. A major difference between the two lessons is how the teacher attends to students' misconceptions about graphs of distance and time. In version 2, the teacher tells students the speed does not change. In version 1, the teacher leaves this misconception to be resolved by students. A second difference is in the nature of questions the teacher asks in each lesson. In version 2, the teacher asks questions of a procedural nature. When students give responses, the teacher makes the connections between the graph and the mathematics involved, and it is never evident whether students are making these connections themselves. In the Father and Son Race lesson version 1, the students are able to make sense of the point of intersection in relation to the story and interpret the graph as indicating a steady rate for the father and son, with the son running faster the whole time and overtaking the father after the point of intersection. There is explicit evidence of students making connections between different aspects of the graph and understanding how the distance and time relate on the graph.

How does the scenario in version 2 affect the level of access each student has to the task compared to the scenario in version 1?

In version 1, students are engaging with the task and with their peers while working on the task. As the teacher circulates around the room, she makes sure to question students and they share their thinking with both her and other students. During the discussion in both the small and whole group, you can notice different students thinking and reasoning about the graph and how their understanding of the graph changes throughout the lesson. The teacher provides access to the students by asking questions and having students engage with each other in a way that promotes their understanding of the mathematics in the graph. In version 2, the over-scaffolding at the beginning of the lesson denies students the opportunity to think and reason. While they may be able to arrive at a correct answer for this specific problem on their own, they may not gain enough understanding of the mathematics to have access to similar problems in the future. In version 1, students make connections and interpret the graph in ways that promote access to solving similar problems about distance and time on graphs. The students, rather than the teacher, access and engage with the mathematics.

Why do high-level demands decline? As seen in version 2, the demands of the task decline when the opportunities for thinking and reasoning are diminished or proceduralized, leaving the students with little opportunity to make sense of connections to the underlying mathematical ideas. In many cases, cognitive demand declines when instruction becomes focused on producing correct answers to specific tasks rather than developing mathematical understanding and relationships that apply to a wide range of future problems, contexts, and mathematical situations.

Summary

In chapter 1, you analyzed the level of cognitive demand of mathematical tasks. We noted that analyzing the *potential* of a task to provide opportunities for thinking and reasoning is an essential first step to promote learning mathematics with deeper understanding as well as to increase access for each and every student. However, even more important is what actually happens during instruction as teachers and students engage with the task during implementation. In this chapter, you explored scenarios in which cognitive demand can be maintained, increased, or decreased during implementation. The IQA Implementation of the Task rubric and Implementation Observation Tool can help identify factors that support students' engagement in cognitively challenging mathematical work and thinking.

Chapter 2 Transition Activity: How Teacher's Questions Impact Implementation

Work within your collaborative team to observe or video record each of you teaching a mathematics lesson. You may want to pair up and observe each other or share videos. For each lesson, do the following.

- Rate the task and indicate what features or criteria from the Potential of the Task rubric contributed to your rating. What level of thinking would the task potentially elicit from students? What products or processes would serve as evidence that students actually engaged in this level of thinking? How does the task provide access for all students?

- Take notes during the observation or as you view the video. In particular, write down all of the questions the teacher asks as students work and during any discussions. Try to capture the exact wording of the teacher's questions. This will be important in chapter 3 when we ask you to categorize the types of questions.
- Reflect on the lesson. Based on your notes, rate the lesson using the Implementation of the Task rubric. Indicate what features or criteria from the rubric seem to characterize the lesson.
- What teacher actions seemed to maintain or enhance students' opportunities for thinking and reasoning? What teacher actions seemed to limit students' opportunities for thinking and reasoning? Be as specific as possible.
- What questions seemed to promote students' opportunities for thinking and reasoning? What questions seemed to elicit procedures, facts, or short responses?

Save the lists of questions, your ratings, and notes or any written reflections from the transition activity, as you will refer to them in chapter 3.

PART 2

Connecting to the *Q* in TQE: Questions and Their Role as Discourse Actions

In part 1 on tasks, you analyzed lessons by considering and rating the task and overall implementation in broad terms. Part 2 connects to the *Q* in the TQE process: "Questions: Facilitate productive questioning during instruction to engage students in the mathematical practices and processes" (Dixon, Nolan, & Adams, 2016, p. 4). In part 2, you will analyze lessons more deeply by identifying specific *discourse actions*, such as asking questions and following up on students' responses that occur during the lesson—in other words, actions teachers use to encourage students to keep talking. You will explore how discourse actions support and encourage students' mathematical work and thinking. As you identify and analyze discourse actions, you will home in on specific instructional practices that impact students' opportunities to engage in thinking and reasoning during mathematics lessons.

As mathematics teachers, once we ask a question and a student provides a response, the next instructional action we make is equally as important as the original question. We can use actions that encourage additional discussion or contributions, or we can use actions that evaluate the student's response and continue on. In chapter 3, we introduce the IQA Teacher's Questions rubric as a tool for analyzing the types of questions mathematics teachers ask during a lesson. Similar to mathematical tasks, different types of questions provide different opportunities for students to engage in rigorous thinking and discourse. In chapter 4, you will explore how teachers follow up on students' contributions. We introduce two IQA rubrics in chapter 4: (1) IQA Teacher's Linking rubric and (2) IQA Teacher's Press rubric. The Teacher's Linking rubric identifies how teachers provide opportunities for students to connect to the mathematical work and thinking of their peers. The Teacher's Press rubric provides a structure to analyze the extent to which teachers elicit further justification and explanation from students. Together these chapters provide guidance as to how your discourse actions support the implementation of mathematical tasks.

CHAPTER 3
Teacher's Questions

In a general sense, teachers' questions control students' learning because they focus students' attention on specific features of the concepts that they explore in class. Moreover, these questions establish and validate students' perceptions about what is important to know to succeed in mathematics class.

—Azita Manouchehri and Douglas A. Lapp

In this chapter, you will explore how to analyze types of questions and differentiate between questions that elicit thinking and reasoning and questions that elicit facts or procedures. By the end of the chapter, you will be able to respond to the following prompts.

- How do different types of questions serve to elicit different types of mathematical thinking and reasoning from students?

- What actions encourage additional discussion or contributions? What actions evaluate the students' responses and then move on?

- What types of questions do I ask during a mathematics lesson, as students work in concurrent small groups and pulled small groups, and as they engage in whole-class discussion?

- In what ways does my questioning provide opportunities for each and every student to engage in the lesson?

Introductory Activities

Let's get started by thinking about different types of questions that teachers may ask during mathematics lessons. In activities 3.1 through 3.4, we ask you and your collaborative team to identify and categorize the types of questions teachers often ask in the classroom and consider the impact this may have on student learning.

Activity 3.1: Identifying Different Types of Questions

Different types of questions elicit different types of responses. We ask you to consider the types of questions teachers ask during instruction.

Engage

In this activity, we revisit both versions of the Father and Son Race lesson.

- Rewatch the Father and Son Race lesson version 1 (page 49) and write down the questions the teacher asks.

- Reread the Father and Son Race lesson version 2 in figure 2.11 (page 50) and make note of the teacher's questions. Imagine that the teacher (in version 2) pulls the class back together following their work in small groups, and the discussion in figure 3.1 occurs. Make note of the questions the teacher asks.
    - How do the teacher's questions in version 1 of the lesson compare to the teacher's questions in version 2 (including the extension)?
    - What types of questions did the teachers ask—how could you categorize them? What purpose did the different types of questions serve in the lesson?

Teacher: Okay, class, let's talk about your stories for the Father and Son Race. Who started here at zero (*pointing to the graph*)? Juan?

Juan: The son.

Teacher: That's right, the son started at zero. And who started at 20? Desiree?

Desiree: The father.

Teacher: Correct. So at the start of the race, the son gave his father a head start. Then they start racing and meet here (*points to the point of intersection*). What do we call this point? Chen?

Chen: The point of intersection.

Teacher: What do we know is the same at this point? Alex?

Alex: They ran the same distance in the same time.

Teacher: Excellent answer! At the point of intersection, they are at the same distance and time. What happens after this point? Sam?

Sam: The son starts running faster and takes the lead.

Teacher: Well, not quite. Remember at the beginning of the lesson, we said that the father and son are running at constant speeds. The son's speed does not change during the race; it remains constant. That's what the straight line tells us. The son is running at a faster constant speed than his father, so picture the son catching up to the father, and then taking the lead. So, class, if you are racing someone and they have a head start, what do you need to do to catch up to them and win the race?

Students: Run faster!

Figure 3.1: Father and Son Race lesson version 2 extension.

Discuss

How do your responses compare with those in your collaborative team? What themes emerged during your discussion? In this section, we present ideas for you to consider.

How do the teacher's questions in version 1 of the lesson compare to the teacher's questions in version 2 (including the extension)?

Just as with tasks, different types of questions prompt different levels of thinking and elicit different levels of responses from students. In version 1 of the lesson, the teacher's questions elicit students' thinking. For example, there are multiple occasions where the teacher asks whether students agree or disagree with one another or prompts students to talk through common misconceptions. In the extension to version 2 of the lesson, the teacher asks mainly procedural or factual questions. Rather than eliciting students' thinking, the teacher provides students with an explanation of the main mathematical ideas.

What types of questions did the teacher ask—how could you categorize them? What purpose did the different types of questions serve in the lesson?

The framework in figure 3.2 (page 60) provides a way to categorize questions. This framework supported the development of the IQA Teacher's Questions rubric, which we will present later in the chapter. Figure 3.2 includes examples of questions from version 1 of the Father and Son Race lesson.

How did your categories compare to the categories in figure 3.2? The figure 3.2 framework, adapted from the work of Jo Boaler and Karin Brodie (2004) and Mary Lou Metz (2007), provides a means of distinguishing between questions that promote mathematical thinking and reasoning (*probing students' thinking, exploring mathematical meanings and relationships,* and *generating discussion* question types) and questions that promote recall or use of mathematical facts, terminology, and procedures (*eliciting procedures or facts*). While both broad categories of questions are necessary in a mathematics lesson, too often the predominance of questions occur at the procedural and factual level (Gokbel & Boston, 2015). For example, in version 2 of the lesson, the teacher asks mostly procedural or factual questions, with two instances of questions that explore mathematical meanings and relationships ("What do we know is the same at this point?" and "What happens after this point?"). The teacher mainly asks questions that require brief answers, so students are limited to providing brief answers. If *eliciting procedures or facts* questions are the primary type of question asked during a mathematics lesson—and during each and every mathematics lesson over time—implementation focuses on the procedural aspects of the mathematics, and therefore students are not likely to see thinking, reasoning, and sense making as important when learning mathematics.

In version 1 of the lesson, however, the majority of questions involve *probing students' thinking, exploring mathematical meanings and relationships,* or *generating discussion* question types—thus providing greater opportunities for students' explanations. The teacher asks questions that provide students the opportunity to offer extended responses. While procedural or factual questions also play an important role in this version of the lesson, the teacher intersperses these questions between questions that elicit thinking and reasoning. Planning and using a balance of cognitively challenging questions and procedural and factual questions throughout a lesson, as needed, helps ensure students engage in high-level mathematical thinking and reasoning.

Another difference between the two versions of the Father and Son Race lesson is the teacher's choice to *ask* versus *tell*. In contrast to version 1 of the lesson, note how the teacher explains the main connections and misconceptions in version 2 rather than eliciting those explanations from students through questioning. For example, the teacher in version 2 summarizes, "The son's speed does not change during the race; it remains constant. That's what the straight line tells us." In version 1, the teacher continues to ask questions until a student (specifically, one who had demonstrated a misconception earlier in the lesson) explains that distance is increasing rather than speed. The teacher uses questions to understand what the student is thinking about the graph. Through questioning, the teacher supports the student in understanding the mathematics, which then allows the student to make connections between the distance and time and how that is represented on the graph. In doing so, the teacher provides scaffolding through her questions.

Framework for Different Types of Questions		
Question Type	**Description**	**Examples From the Father and Son Race Lesson Version 1**
Probing students' thinking	• Clarifies student thinking • Enables students to elaborate their own thinking for their own benefit and for the class	• So what are you doing? • What are we talking about with accelerating? • Time is standing still? Try to explain. • So what is your thinking about this now?
Exploring mathematical meanings and relationships	• Points to underlying mathematical relationships and meanings • Makes links between mathematical ideas	• What was happening at the start of the race? • Where are they when the race starts? • What in the graph tells you that? • What is happening with the speed? • What's happening at B? • Where are they at C? • What does the "up" mean? • Is there a point during this race when the son was running at a different rate?
Generating discussion	• Enables other members of the class to contribute and comment on ideas under discussion	• Do you agree? • So you agree it is faster? • What does she mean by that? • Why does she say "no"? • Agree or disagree? Why?
Eliciting procedures or facts	• Elicits a mathematical fact, procedure, or definition • Requests the result (only) of a mathematical procedure • Requests units or terminology • Requires a yes-or-no or single-word answer	• At four what? • Zero what? • What do we call what is happening at A?
Inquiring about other mathematical topics	• Relates to teaching and learning mathematics, but does not directly relate to the task or mathematical ideas of focus for the lesson	General examples (not from the lesson): • What else could you graph and find the point of intersection? • How else could you label this graph?
Asking nonmathematical questions	• Does not relate to teaching and learning mathematics • May relate to the context of the lesson or task	General examples (not from the lesson): • Do you want to use graph paper? • Who has competed in a race in track?

Source: Adapted from Boston, 2017.

Figure 3.2: Framework for considering different types of questions.

Having an awareness of different types of questions can enable you to:

- Plan questions that both promote mathematical thinking and reasoning and elicit mathematical facts *prior to* the lesson
- Ask a balance of different types of questions throughout a lesson
- Use questions as a tool for diagnosing students' understandings and misconceptions
- Support students' overall engagement in thinking and reasoning (in other words, support the implementation of the task at a level 3 or 4)

In activities 3.2 and 3.3, you will have the opportunity to identify and create questions in each of the categories in figure 3.2.

Activity 3.2: Sorting Questions

Different types of questions elicit different types of responses. We ask you to consider examples of the different types of questions teachers ask during instruction.

Engage

This activity helps you to clarify the different types of questions.

- Print the reproducible for figure 3.3 (page 63), and cut out the questions.
- Using the template in figure 3.3, sort the questions by category. Provide a rationale for your sorting choice. Discuss your question sorting with your collaborative team before continuing on to the activity 3.2 discussion.

Discuss

How do your responses compare with those in your collaborative team? What themes emerged during your discussion? In this section, we present ideas for you to consider.

Using the template in figure 3.3 (page 63), sort the questions into each category. Provide a rationale for your choice of categories.

One possible sorting of the questions in figure 3.3 is provided in appendix C (page 139). Sometimes, the same question might serve different purposes depending on the context in which the question is asked. A teacher might ask a question to probe students' thinking in order to generate discussion among students. A single question might also serve multiple purposes. For example, a question to probe students' thinking might also prompt students to make mathematical connections. For our purposes, we consider the categories of *probing students' thinking, exploring mathematical meanings and relationships*, and *generating discussion* as collectively supporting students' engagement in cognitively challenging mathematical work and thinking. *Procedural and factual* questions serve the important function of eliciting mathematical procedures, facts, or definitions, but they do not provide students opportunities for thinking and reasoning.

In activity 3.3, you will have the chance to create questions in each category.

Probing Students' Thinking	**Exploring Mathematical Meanings and Relationships**	**Generating Discussion**
• Clarifies student thinking • Enables students to elaborate their own thinking for their own benefit and for the class	• Points to underlying mathematical relationships and meanings • Makes links between mathematical ideas	• Enables other members of the class to contribute and comment on ideas under discussion
Eliciting Procedures or Facts	**Inquiring About Other Mathematical Topics**	**Asking Nonmathematical Questions**
• Elicits a mathematical fact, procedure, or definition • Requests the result (only) of a mathematical procedure • Requests units or terminology • Requires a yes-or-no or single-word answer	• Relates to teaching and learning mathematics, but does not directly relate to the task or mathematical ideas of focus for the lesson	• Does not relate to teaching and learning mathematics • May relate to the context of the lesson or task

Teacher's Questions

How did you come up with your answer?	How does your table relate to your graph?	Who agrees with what Ryan said? Why do you agree?	What is a word problem that could be modeled by this expression?
How did you determine the scale for your graph?	Could you use that formula to find the volume of a different object?	What is the square root of 16?	Which problem was the most difficult for you to solve?
Why did you use that algorithm to solve the problem?	What is staying the same in the graph? Why is it staying the same?	What is the denominator?	How could you describe Cameron's solution process in your own words?
Explain to me how you got that product.	What did Keisha say?	What is 6 × 7?	Who has ever been skiing?
What does x represent in the equation?	What else did you notice about the graph of the function?	Is the graph linear or quadratic?	What is the definition of a quadrilateral?

Figure 3.3: Template and questions for activity 3.2.

*Visit **go.SolutionTree.com/mathematics** for a free reproducible version of this figure.*

Activity 3.3: Creating Questions

Different types of questions elicit different types of responses. We ask you to create questions you could use during a mathematics lesson.

Engage

In this activity, we ask you to create questions based on student work.

- Imagine that you observe students in your classroom in the act of producing the grade 4 student work for the 26 Divided by 4 task (figure 2.5, page 40) and the grade 6 student work for the Area of a Trapezoid task (figure 2.6, page 41) during a mathematics lesson as they work in small groups. Use the template in figure 3.3 to write two to three questions in each category that you could ask as you circulate among the small groups while they are working.
- Discuss the questions you create with your collaborative team.

Discuss

How do your responses compare with those in your collaborative team? What themes emerged during your discussion? In this section, we present ideas for you to consider.

Use the template in figure 3.3 to write two to three questions in each category that you could ask as you circulate among the small groups.

Some examples of questions in each category include those provided in figure 3.4 (page 65).

Compare the *eliciting procedures or facts* questions (darkly shaded) to the questions in the collective categories of *probing students' thinking, exploring mathematical meanings and relationships,* and *generating discussion* (lightly shaded). Questions that elicit procedures or facts produce one brief correct answer, while the lightly shaded categories elicit deeper mathematical work and thinking. In planning and teaching mathematics lessons, be sure to incorporate questions that fall in the lightly shaded categories in order to understand student thinking.

Asking nonmathematical questions and *inquiring about other mathematical topics* questions may serve the important purpose of ensuring students understand the context of the problem. One crucial aspect of providing access to each and every student in the lesson is making sure students understand the context of what you are asking and thus have an entry point into the task (Jackson, Shahan, Gibbons, & Cobb, 2012). It is important to think about what questions you can ask while introducing a task so that the context of the lesson does not serve as a barrier for students. Plan questions prior to the lesson that will ensure students understand the context of the problem and what the task is asking them to do. The Father and Son Race task provides an example of the need for questions related to the context of the task. Students may need help understanding that the father and son are competing in a running race against each other, rather than as a team, or in a "race" to see who finishes first in competition. Also, some cultures or families may consider a son winning a race against his father to be disrespectful. While asking if students understand the context of the race is nonmathematical in nature, these questions are necessary to make sure each and every student understands the context represented by the graph and has access to the task.

Question Type	26 Divided by 4	Area of a Trapezoid
Probing students' thinking	• Figure 2.5 A and D: How did you decide on that context or story? What is it about the situation that causes the answer to be 7?	• Why did you decide to partition your trapezoid that way? • How do you know how to find the dimensions of the triangles?
Exploring mathematical meanings and relationships	• Figure 2.5 C: Does "7" answer the question, How many boxes can he fill? How might you revise the question so that 7 is the answer to the question? • Figure 2.5 A, C, and D: Can you change the situation so that the answer is 6? How are the situations the same as or different from what you have here?	• How does your expression relate to the diagram? • Can you find another way to partition the trapezoid?
Generating discussion	• Figure 2.5 B: Do you all agree there is not a situation in which the answer is 7?	• How did you both get the same result if you partitioned the trapezoid differently? • How are your strategies (or expressions) the same or different?
Eliciting procedures or facts	• What is 26 divided by 4? • Figure 2.5 C: How many boxes would we need for 31 muffins? How many boxes would we fill for 31 muffins?	• What is another way to write $½b_1h + ½b_2h$? • What is the formula for the area of a triangle? • What is the definition of a trapezoid?

Inquiring about other mathematical topics	• Have you written mathematics stories before?	• What are some examples of other types of quadrilaterals? • Where do you notice trapezoids in the real world?
Asking nonmathematical questions	• Figure 2.5 C: What is your favorite type of muffin?	• Is it difficult to draw a trapezoid? • Did you use complete sentences in your explanation?

Figure 3.4: Questions for 26 Divided by 4 and Area of a Trapezoid tasks.

Activity 3.4: Revisiting the Chapter 2 Transition Activity—How Teacher's Questions Impact Implementation

In the chapter 2 transition activity (page 52), you analyzed the task, implementation, and questions that occurred in a mathematics lesson. In the next activity we ask you to discuss your ideas with your collaborative team.

Engage

With your collaborative team, do the following.

- Discuss your rating for the task and indicate what features or criteria from the Potential of the Task rubric contributed to your rating. Describe whether the task provides access for all students and justify your reason.

- Discuss your rating for the implementation. Indicate what features or criteria from the Implementation of the Task rubric seem to characterize the lesson. Indicate what teacher actions from the Implementation Observation Tool seemed to maintain, increase, or decrease students' opportunities for thinking and reasoning.

- For the questions you recorded, identify examples of each category.

    □ What questions seemed to promote students' opportunities for thinking and reasoning? What questions seemed to elicit procedures, facts, or short responses?

    □ Were there any major question categories that were missing? If so, what questions could the teacher have used, and how might those questions promote opportunities for students' thinking and reasoning?

Discuss

How do your responses compare with those in your collaborative team? What themes emerged during your discussion? In this section, we present ideas for you to consider.

What questions seemed to promote students' opportunities for thinking and reasoning? What questions seemed to elicit procedures, facts, or short responses?

The type of questions asked during a lesson impacts the overall implementation of the task. *Probing students' thinking, exploring mathematical meanings and relationships*, and *generating discussion* question types can all support the implementation of a cognitively challenging task. For example, in the Father and Son Race lesson version 1, the teacher continues to ask questions to support students' understanding of the task but also to elicit misconceptions, rather than guiding them down a pathway or providing a correct solution. In fact, the teacher is purposeful in asking broad questions to small groups (*probing students' thinking* and *generating discussion*) and then asking more pointed questions during the whole-group discussion (*exploring mathematical meanings and relationships*) so the entire class is able to wrestle with the idea of increasing rate of change versus constant rate of change.

Questions can serve to increase cognitive demand from a Potential of the Task rating of level 3 to an Implementation of the Task rating of level 4 by probing students' thinking for explanations and reasoning. For example, in the Leftover Pizza lesson version 1 in chapter 2 (page 30), the teacher asked questions that required explanations of students' thinking and reasoning, even though an explanation was not specifically indicated in the task. However, for a task that rated a level 1 or 2 for Potential of the Task to have an Implementation of the Task rating of level 3 or 4 would require more than probing students' thinking for explanation. With memorization or procedural tasks, if the only thing to discuss is the procedure, answer, or memorized knowledge, probing students' thinking for an explanation may only elicit a series of steps (how students solved the problem) rather than students' thinking and reasoning (why the steps make sense, or connections to big mathematical ideas). Questions in the category of *exploring mathematical meanings and relationships* would be necessary to promote the type of thinking, reasoning, or connections necessary for a level 3 or 4 for Implementation of the Task.

Questions can also serve to provide entry for students into the work of the problem. A teacher's questions can scaffold and support students to gain access to the task as written. Rather than breaking down a challenging task into smaller subparts that might lower the demands of the task and take away or over-scaffold students' thinking, teachers can use questions to provide scaffolding *just in time* (Dixon, Brooks, & Carli, 2018)—that is, at the moment a student needs it.

Conversely, a predominance of *eliciting procedures or facts* questions can serve to reduce the demands of a cognitively challenging task during implementation. For example, in figures 2.1 (page 31) and 3.1 (page 58), the teacher's questions leave room for only short responses from students, and the main mathematical ideas are provided by each of the teachers rather than the students.

Reflect on the Father and Son Race lesson version 1 (page 49). In this version of the lesson, the teacher provides scaffolding just in time rather than just in case (Dixon, Brooks, & Carli, 2018). If the teacher had approached a group early in the exploration process and asked the students, "What is the slope for the son's race?" and then, "What does a constant slope tell you?" the teacher would have been providing scaffolding just in case the students needed it. Rather, the teacher allows students to engage in productive struggle by asking the students what they are doing. Even though the students are not correct in their responses, she lets them persist in the struggle because it is too early to provide scaffolding. Later in the lesson, the teacher asks specifically about what is happening with the father's and son's speeds. This is to provide a scaffold to focus on the slope of the graph. This scaffolding is provided just in time as the struggle becomes unproductive.

Were there any major question categories that were missing? If so, what questions could the teacher have used and how might those questions promote opportunities for students' thinking and reasoning?

If you find that questions from the categories of *probing students' thinking, generating discussion*, or *exploring mathematical meanings and relationships* were missing from the observed lesson, use the examples in figure 3.2 (page 60) and figure 3.4 (page 64) to develop lesson-specific questions that could have been used to support students' thinking during instruction. The process of creating questions within your collaborative team will help to make asking questions a seamless part of your discourse actions.

Questions play a pivotal role in overall implementation and heavily impact the rating on the Implementation of the Task rubric. Because we acknowledge with many others that teachers' questioning is so important (NCTM, 2014; Sztajn, Confrey, Wilson, & Edgington, 2012), we provide a separate rubric within the IQA focused on this aspect of teachers' instructional quality. This rubric is aptly named the IQA Teacher's Questions rubric.

The IQA Teacher's Questions Rubric

The IQA Teacher's Questions rubric (figure 3.5) provides guidance for developing a rating of the level of questions a teacher asks during a lesson. In this rubric, consistent use of the *probing students' thinking, exploring mathematical meanings and relationships*, and *generating discussion* question types is required to rate a level 4, even if other questions asked throughout the lesson are procedural or factual.

	IQA Teacher's Questions Rubric
4	The teacher consistently asks academically relevant questions that provide opportunities for students to elaborate and explain their mathematical work and thinking (*probing students' thinking* and *generating discussion*); identify and describe the important mathematical ideas in the lesson; or make connections among ideas, representations, or strategies (*exploring mathematical meanings and relationships*).
3	At least three times during the lesson, the teacher asks academically relevant questions that provide opportunities for students to elaborate and explain their mathematical work and thinking (*probing students' thinking* and *generating discussion*); identify and describe the important mathematical ideas in the lesson; or make connections among ideas, representations, or strategies (*exploring mathematical meanings and relationships*).
2	The teacher makes limited, superficial, trivial, or formulaic efforts to ask academically relevant questions that provide opportunities for students to elaborate and explain their mathematical work and thinking (*probing students' thinking* and *generating discussion*); identify and describe the important mathematical ideas in the lesson; or make connections among ideas, representations, or strategies (*exploring mathematical meanings and relationships*). For example, the teacher asks every student the same question or set of questions; there are one or two instances of strong questions; or the teacher asks the same strong question multiple times.
1	The teacher asks procedural or factual questions that elicit mathematical facts or procedures or require brief, single-word responses (*eliciting procedures or facts*).
0	The teacher does not ask questions during the lesson, or the teacher's questions are not relevant to the mathematics in the lesson (*inquiring about other mathematical topics* or *asking nonmathematical questions*).

Source: Adapted from Boston, 2017.

Figure 3.5: IQA Teacher's Questions rubric.

Use the Teacher's Questions rubric with the lessons for the chapter 2 transition activity (page 52). When rating lessons using the Teacher's Questions rubric during an observation or from video, note examples of *probing students' thinking, exploring mathematical meanings and relationships*, and *generating discussion* question types as a set (the lightly shaded categories in figure 3.4, page 65). When deciding on a rating, consider the following questions.

- **Level 4:** Did the teacher ask *probing students' thinking, exploring mathematical meanings and relationships*, and *generating discussion* types of questions consistently throughout the lesson?
- **Level 3:** Were there at least three good examples (but not a consistent use) of *probing students' thinking, exploring mathematical meanings and relationships*, and *generating discussion* types of questions asked throughout the lesson? Were there at least three *unique* questions in the categories of *probing students' thinking, exploring mathematical meanings and relationships*, and *generating discussion*?
- **Level 2:** Did the teacher ask the same questions or sets of questions repeatedly? Were there fewer than three instances of questions in the categories of *probing students' thinking, exploring mathematical meanings and relationships*, and *generating discussion*?
- **Level 1:** Was the lesson characterized by *procedural and factual* questions?

Discuss your ratings with your collaborative team before moving on to the next activity in the following section. In this activity, we ask you to practice using the IQA Teacher's Questions rubric.

Application Activities

It is valuable to consider the types of questions asked during a mathematics lesson. In the next activity you will have the opportunity to rate the questions posed by the teacher.

Activity 3.5: Rating the Teacher's Questions in the 26 Divided by 4 Lesson

In activity 3.5, you will have the chance to rate a lesson using the IQA Teacher's Questions rubric.

Engage

For this activity you will need the Teacher's Questions rubric (figure 3.5, page 67) to rate the 26 Divided by 4 lesson.

- Watch the 26 Divided by 4 lesson (grade 4). As you view the lesson, write down the questions the teacher asks during the lesson.
- Use the IQA Teacher's Questions rubric (figure 3.5, page 67) to rate the questioning that occurs during the lesson. You may want to use figure 3.4 to identify questions that are *probing students' thinking, exploring mathematical meanings and relationships*, and *generating discussion; eliciting procedures or facts;* or *asking nonmathematical questions* or *inquiring about other mathematical topics*. Discuss your rating and examples of questions with your collaborative team.

26 Divided by 4 Lesson:
www.SolutionTree.com/Interpreting_the_Remainder_in_Word_Problems

Discuss

How do your responses compare with those in your collaborative team? What themes emerged during your discussion? In this section, we present ideas for you to consider.

Use the IQA Teacher's Questions rubric to rate the questioning that occurs during the lesson.

Using the Teacher's Questions rubric, the questioning in this lesson would rate a level 4. We provide examples of the different types of questions the teacher asks during the lesson in figure 3.6. The teacher consistently asks questions that provide opportunities for students to elaborate and explain their mathematical thinking. Procedural and factual questions are interspersed throughout the lesson as needed.

Question Type	Examples of Questions Asked During the 26 Divided by 4 Lesson
Probing students' thinking	• So what does that mean? • What are you doing? • So what could the question be?
Exploring mathematical meanings and relationships	• How can we write a word problem so the answer is 7? • So how can we change this to make sure we would use 7 packages? • Okay, so how can I write this word problem with my 26 pencils so I use up all the pencils and my answer is 7?
Generating discussion	• So she is saying how many pencils would he . . . and then you got a little confused, so what do you guys think? • How many packages are there? Try that. And then talk about what answer you might get in your group. • Do you want to help her out?
Eliciting procedures or facts	• Alex has 26 books. She wants to divide them into 4 equal groups, so how many books will she put in each group? • So how many pencils are you going to put in each package? • So how many packages would I need?
Inquiring about other mathematical topics	• I want you to write a word problem for 26 divided by 4, where the answer would need to be 7. Now that's a little strange, isn't it? Why?
Asking nonmathematical questions	

Figure 3.6: Examples of questions from the 26 Divided by 4 lesson.

Note how the teacher uses the *inquiring about other mathematical topics* questions (figure 3.6, page 69) in the launch of the task (timestamp 0:23) to make sure students understand what it means to create a story problem with an answer of 7. She first asks what is different about the task, and a student explains there is not a whole number you can multiply by 7 to get 26. The teacher then asks a *probing students' thinking* question ("So what does that mean?" [0:35]) to get students thinking about the structure of the problem, without providing too much guidance or taking away students' opportunities for thinking and reasoning. With a small group (2:05 to 2:59) and during the whole-group discussion (3:10 to 6:15), she uses questions to help students understand the difference between writing a story problem where the answer would be 6 versus 7 and how to think about the remainder to give a solution of 7. At the end of the lesson, once a correct story problem is shared in the whole-class discussion, the teacher uses questions to make sure other students understand why it works ("So what does she mean?" [5:50]). By asking questions and including students' voices in both the small- and whole-group discussions, the teacher promotes students' access and engagement throughout the lesson.

As indicated by the 26 Divided by 4 lesson, different types of questions serve different purposes within a lesson. Questions in the categories of *probing students' thinking, exploring mathematical meanings and relationships*, and *generating discussion* serve to maintain students' engagement in thinking, reasoning, and problem solving. Consistent use of these types of questions throughout a lesson rates a level 4 on the IQA Teacher's Questions rubric, even if other types of questions (*eliciting procedures or facts, inquiring about other mathematical topics*, and *asking nonmathematical questions*) are also present. Similarly, using at least three questions from these categories rates a level 3, and the occurrence of at least one question from these categories (or the same question asked repeatedly) rates a level 2, even if many other types of questions are asked. *Procedural or factual* questions may help a teacher assess students' knowledge and may help students recall or retrieve facts or procedures as they work on a task. Used in conjunction with the other types of questions, *procedural or factual* questions are often necessary and useful. However, if teachers ask only procedural or factual questions during a lesson, students would not be held accountable for thinking and reasoning—only for reproducing procedural or factual knowledge—and the lesson would rate a level 1 on the Teacher's Questions rubric. *Inquiring about other mathematical topics* or *asking nonmathematical questions* may serve the purpose of familiarizing students with the context or requirements of the task, thereby promoting entry and access to the task for each and every student. To promote students' engagement, however, such questions would need to be followed with *probing students' thinking, exploring mathematical meanings and relationships*, and *generating discussion* question types throughout the lesson.

Summary

The questions you ask during a mathematics lesson can support students to explain their ideas, hear the ideas of their peers, develop new mathematical insights and connections, and talk through misconceptions. The tools provided in this chapter, particularly the question types in figure 3.4 (page 65) and the IQA Teacher's Questions rubric in figure 3.5 (page 67), can support your efforts to plan and use the types of questions that elicit students' thinking and reasoning.

While *probing students' thinking* or *generating discussion* questions may occur naturally in response to students' work and thinking as a lesson unfolds, questions that prompt students to explore mathematical meanings and relationships may be more difficult to develop in the moment during a lesson. Planning such questions in advance can equip you to support students' thinking and reasoning during a lesson. Well-crafted questions support your efforts to *ask* versus *tell* by providing an opportunity to hear students explain mathematical ideas rather than having you explain ideas to them. By asking questions, you can shift who is doing the talking and thinking in the mathematics classroom.

Questions are also very useful within the formative assessment process, and provide the opportunity for you to assess students' mathematical understanding. While beyond the scope of this book, it is also important to examine the recipients of the questions. Do certain students get more thought-provoking questions while other students are limited to questions that require factual or procedural responses? Questions can be used to understand students' thinking and to help you decide when to support students in making connections by providing scaffolding just in time. And, as you have read in this chapter, questions can also be used to assess your instructional quality.

Chapter 3 Transition Activity: Teacher's Questions and Follow-Up

Work within your collaborative team to observe or video record each of you teaching a mathematics lesson. You may want to pair up and observe each other or share videos. For each lesson, do the following.

- Obtain a copy of the task and take notes during the observation or as you view the video. Write down examples of the questions the teacher asks as students work and during any discussions. Also, note how the teacher follows up on students' responses to the questions.
- Rate the Potential of the Task and the Implementation of the Task using the rubrics. How do the task and implementation provide access for all students?
- Rate the Teacher's Questions using the rubric. Try to capture the exact wording of the teacher's questions. Provide examples of questions that seemed to promote students' opportunities for thinking and reasoning. Provide examples of questions that seemed to elicit procedures, facts, or short responses.
- In what ways did the teacher follow up on students' responses? Did the teacher ask students to provide more extended explanations or to build on the ideas of their peers? Did the teacher provide an evaluation of the students' responses? Try to capture the exact wording of the teacher's follow-up. This will be important in chapter 4 when we ask you to categorize the ways in which the teacher follows up on students' contributions.

Save your responses, ratings, and notes or any written reflections from the transition activity, as you will refer to them in chapter 4.

CHAPTER 4
Teacher's Linking and Teacher's Press

One major way to improve mathematics instruction and learning is to help teachers understand the mathematical thought processes of their students.

—Elizabeth Fennema, Thomas P. Carpenter, Megan L. Franke, Linda Levi, Victoria R. Jacobs, and Susan B. Empson

In this chapter, you will explore how to enhance students' contributions to mathematical discussions. During classroom conversations, the same question can sometimes elicit very different responses from students. You will consider how to follow up on students' contributions in ways that elicit deeper explanations and justifications and prompt students to engage with the ideas of their classmates. You will be able to answer the following questions.

- During mathematical discussions, how are students encouraged to build from and connect to each other's ideas in a way that makes their ideas clear to the rest of the class?

- In what ways can a teacher follow up on students' contributions to encourage deeper explanations and justifications about the mathematics?

Introductory Activities

Let's get started by considering different ways teachers follow up on students' contributions during mathematical discussions. In activities 4.1 through 4.3 we ask you and your collaborative team to think about ways teachers make links between individual students' ideas and press students for additional information in their thinking.

Activity 4.1: Following Up on Students' Contributions—Teacher's Linking

How do teachers follow up on students' responses to make their ideas public to the rest of the classroom community? In activity 4.1, you will examine a teacher's actions during a whole-group discussion more closely.

Engage

In chapter 2, you watched the Leftover Pizza lesson version 1 (page 30) and considered the overall Implementation of the Task. Recall that the Potential of the Task rated a level 3, because the task provided opportunities for reasoning and sense making but did not explicitly prompt for an explanation. In version 1 of the Leftover Pizza lesson, the Implementation of the Task rated a level 4 because throughout the lesson, students engaged in sense making and provided explanations of their thinking and reasoning.

- Watch the Leftover Pizza lesson version 1 (page 30) again.
- During the small-group and whole-class discussions, what does the teacher say following a student's response to encourage students to relate to and build on each other's ideas? Write down examples of when the teacher follows up on students' contributions. Discuss your examples with your collaborative team before moving on to the activity 4.1 discussion.

Discuss

How do your responses compare with those in your collaborative team? What themes emerged during your discussion? In this section, we present ideas for you to consider.

During the small-group and whole-class discussions, what does the teacher say following a student's response to encourage students to relate to and build on each other's ideas?

Students are attempting to determine the number of servings that can be made from $4\frac{2}{3}$ pizzas if each serving is $\frac{2}{3}$ of a pizza and all the pizza is used. A common error that is not productive is to assume that only four servings can be made. A student in one of the small groups the teacher visits makes this error (1:50):

> **Teacher:** What has she figured out here?
>
> **Student:** That you can make four servings.
>
> **Teacher:** Do you agree that you can make four servings?
>
> **Student:** Yes.

The teacher then asks, "So what are you going to do with this piece, just throw it away? . . . You're going to throw it away?" and students realize, "No, you're going to freeze it!" The teacher's questions in this exchange—and her subtle shifting of the manipulative to bring two unused $\frac{1}{3}$ pieces together—serve the specific purpose of engaging students with an idea offered by one of their peers and, in this example, help students resolve a misconception.

Students eventually arrive at two answers: $7\frac{1}{6}$ and $7\frac{1}{4}$. Students use models to determine that seven servings of pizza can be placed into freezer bags, and then they wrestle with how to name the portion of pizza that remains: $\frac{1}{6}$ of a pizza, but $\frac{1}{4}$ of a serving. The whole-class discussion begins with the teacher identifying two of the results she heard as students worked in small groups. She chooses to discuss these specific answers because one of the responses is the correct answer and the other is an important common error in fraction division—and is the reason this task is so useful. The teacher asks for students to defend their answers (2:48): "I've heard two answers that we're going to discuss. I heard a group say they found $7\frac{1}{4}$ servings. I heard another group say they found $7\frac{1}{6}$ servings. . . . We need people to defend their answers."

This discourse action provides an opportunity for students to hear the reasoning of their peers. A student volunteers to explain the result of $7\frac{1}{6}$. He describes his strategy and model, and identifies the remaining piece of pizza as $\frac{1}{6}$ of a pizza. The teacher responds by identifying the aspects of the explanation and model that are valid (seven servings), and then asks the class whether they agree with the $\frac{1}{6}$ (5:08):

> *So we see how he got seven servings; do we see how he got ⅙? So do we agree with that? I'm hearing yeses and noes. . . . Who doesn't agree with him?*

This discourse action provides an opportunity for students to engage with ideas offered by their peers. A student responds with an explanation that is still focused on the portion of the whole pizza (5:19): "I disagree with him because it was six parts when it was a whole." However, this explanation based on the whole pizza does not support students in making sense of fraction division. The teacher then redirects her questions to a student who she knew could provide a justification for naming the leftover piece of pizza as ¼ of a serving (5:33): "I remember you talking about servings before. You want to talk to us about it?" In doing so, the teacher provides an opportunity for all students in the class to hear the contrasting "names" for the portion of pizza that is left over: ⅙ of a pizza and ¼ of a serving. Following the student's explanation of why the remaining piece is ¼ of a serving, she then returns to the two previous students to determine whether they now understand the correct result of 7¼ servings, based on the explanations of their peers:

> **Teacher:** *Are you able to make sense of what he said? (5:49)*
>
> **Teacher:** *So now where is your thinking? (5:57)*

Through these discourse actions, the teacher prompts students to connect the ideas and positions of their peers so that these ideas are public to the classroom community, with the purpose of supporting each and every student to understand the mathematical ideas in the lesson. In the IQA Toolkit, we describe these types of actions as *teacher's linking* (Boston, 2012, based on the work of Michaels, O'Connor, Hall, & Resnick, 2010). Table 4.1 (page 76) provides examples of teacher's linking.

Teacher's linking keeps students' ideas at the center of mathematical discussions, highlights students' contributions as important for the learning of the entire class, and provides students with additional opportunities to hear and engage with the mathematical thinking of their peers.

In the *Making Sense of Mathematics for Teaching* series (for example, in Dixon, Nolan, Adams, Tobias, & Barmoha, 2016), the authors describe three essential classroom norms.

1. Explain and justify solutions.
2. Make sense of each other's solutions.
3. Say when you don't understand or when you don't agree.

Teacher's linking supports students to engage in the second and third norms, with the purpose of keeping ideas public in the classroom space so that more students have the opportunity to make sense of the mathematical ideas.

In what ways can teachers support students to engage in the first norm of explaining and justifying their solutions? In activity 4.2, you will explore a second set of discourse actions that can be used for this purpose identified by the IQA Toolkit.

Table 4.1: Examples of Teacher's Linking

Teacher's Linking	Description	Examples
Revoicing students' contributions	Revoicing involves repeating or rephrasing students' responses with intentional emphasis or clarity. This includes inserting mathematical vocabulary, rephrasing parts of the response as a question to prompt students to offer additional explanations, and emphasizing important words or ideas to make them more salient or mark them as important to the entire class.	• Student: "I plussed 4 and 7." Teacher: "What I am hearing you say is, you *added* 4 and 7?" • Student: "12 is the highest point on the graph." Teacher: "So you found point (7, 12) to be the maximum?"
Prompting students to take up the ideas of their peers	When prompting students to take up the ideas of other students, a teacher encourages students to build on, analyze, support, or disagree with the mathematical work and thinking of others.	• "Who can add on to what Taylor is saying?" • "How does what you are saying relate to what Jay is saying?" • "Who agrees and who disagrees with what Alejandro said?"
Focusing attention on students' explanations	Focusing on students' explanations can be seen as keeping everyone together in the class to bring attention to specific contributions. Focusing actions include asking students to repeat responses or marking a response as important. Using focusing actions, the teacher is asking students to revoice an idea from their peers.	• "Can you say that again so that everyone can hear you?" • "Chris just said something very important; can someone restate it in her or his own words?" • "Can you repeat what Peyton said in your own words?"

Source: Boston, 2012; Michaels et al., 2010.

Visit **go.SolutionTree.com/mathematics** *for a free reproducible version of this table.*

Activity 4.2: Following Up on Students' Contributions—Teacher's Press

In activity 4.2, you will consider how the teacher prompts first-grade students to provide explanations of their thinking.

Engage

In the following Change Unknown lesson, students explore the following task: "Stefan has 7 stickers. How many more stickers does he need to have 15 stickers altogether?" This task is challenging for many first-grade students due to the size of the numbers and to the *change unknown* format of the question, as students are asked to find the amount of change (how many more stickers) rather than the final amount of stickers (15 stickers).

- Begin by watching the Change Unknown lesson. As you watch the lesson, consider what the teacher says following students' contributions to encourage students to provide deeper explanations and justifications.
- Write down examples of how the teacher follows up on students' contributions to encourage students to provide deeper explanations and justifications.

Discuss your examples with your collaborative team before moving on to the activity 4.2 discussion.

Change Unknown Lesson:
www.SolutionTree.com/Solving_a_Word_Problem
_Where_the_Change_Is_Unknown

Discuss

How do your responses compare with those in your collaborative team? What themes emerged during your discussion? In this section, we present ideas for you to consider.

What does the teacher say following students' contributions to encourage students to provide deeper explanations and justifications?

During the whole-group discussion (beginning at timestamp 1:20), three students describe their strategies, and one student describes the make-a-ten strategy that a classmate uses. Following students' initial responses, the teacher often asks students to say more or to clarify in some way. For example, the first student (1:31) ends the explanation with "So that is 8," and the teacher follows up with (1:41), "What's 8?" Other examples of prompting students to provide greater detail include:

- (2:36) Fifteen, how do you know it equaled 15?
- (3:24) You put 7 + 3 . . . (3:30) and then what did you need to do?
- (3:52) And then what?

The teacher also prompts students for deeper explanations. When a student describes how he knows 7 + 8 is the same as 8 + 7 (namely, that he understands the commutative property), the teacher asks (1:58), "Why can we do that?" Toward the end of the discussion, the teacher follows up a student's response by asking (3:59), "So how many more stickers did Stefan need?" to have the student clarify that 8 more stickers are needed. Through these discourse actions, the teacher prompts students to expand and deepen their explanations. In the IQA Toolkit, we describe these types of actions as *teacher's press* (Boston, 2012). Table 4.2 (page 78) provides examples of different types of teacher's press.

Table 4.2: Examples of Teacher's Press

Teacher's Press	Description	Examples
Prompting students to explain their thinking	The teacher prompts students to provide thorough explanations of their ideas.	• "And then what did you need to do?" • "What do you mean by rise over run?"
Eliciting students to provide justifications	The teacher asks students to provide appropriate justification for ideas, claims, and conjectures.	• "Why can we do that?" • "How do you know your strategy will always work?"
Asking students to validate mathematical accuracy	The teacher compels students to validate the accuracy of their mathematical computations and facts, maintaining an expectation for accurate knowledge.	• "Fifteen, how do you know it equaled 15?" • "How did you define *trapezoid*?"

Source: Boston, 2012; Michaels et al., 2010.

Visit **go.SolutionTree.com/mathematics** *for a free reproducible version of this table.*

In the lessons in activities 4.1 and 4.2, the teacher uses discourse actions to generate and extend the discussion among students. In the Leftover Pizza lesson version 1 (page 30), you identified teacher's linking actions that provide opportunities for students to engage with the ideas of their peers. Throughout the lesson, the teacher also makes several instances of press. For example, she consistently prompts the first student to say more about his diagram and strategy (3:12 to 5:06). In the Change Unknown lesson, you identified instances of teacher's press that prompt students to provide deeper explanations. In this lesson, the teacher also makes two important linking actions. The first is revoicing, when the teacher says (1:50), "So I'm hearing you say that you knew 8 plus 7 equals 15." In the second example, the teacher prompts students to take up the ideas of their peers when she says (3:38), "How did he use 'making a 10' to solve that problem?"

Traditionally, discussions in mathematics classrooms have followed a pattern of Initiate-Response-Evaluate (IRE; Mehan, 1979): the teacher *initiates* (I) a question, a student provides a *response* (R), and the teacher *evaluates* (E) whether the response is correct or incorrect. In the IRE pattern, typically the teacher only asks one student to provide his or her mathematical ideas or strategy, and then the teacher moves on to the next question. Also, the teacher, rather than other students, serves as the mathematical authority by evaluating the correctness of the response (Boaler & Staples, 2008). When a teacher follows up by linking or pressing, the *E* is replaced by questions or prompts that foster additional contributions from students and contributions from additional students.

A teacher can create opportunities for student-to-student discourse through the use of linking. By highlighting students' contributions, linking also positions students as authors of ideas (Wagner & Herbel-Eisenmann, 2009) and assigns students competence as mathematicians (Cohen & Lotan, 2014). Through linking, the teacher asks students to talk about someone else's ideas or strategies and make those ideas public to the class. Through press, the teacher asks students to explain, justify, or validate their own ideas and strategies. Teacher's press often involves teacher-student discourse. The teacher's intent is to ask questions to specific students; however, teacher's press provides other students the opportunity to hear the reasoning and justification of their peers.

Next, you will have the chance to identify teacher's linking and teacher's press in the lessons you analyzed for the chapter 3 transition activity.

Activity 4.3: Revisiting the Chapter 3 Transition Activity—Teacher's Questions and Follow-Up

In the chapter 3 transition activity (page 71), you analyzed the task, implementation, questions, and teacher follow-up that occurred in a mathematics lesson. In the next activity, we ask you to discuss your ideas with your collaborative team.

Engage

With your collaborative team, do the following.

- Discuss your ratings for Potential of the Task, Implementation of the Task, and Teacher's Questions. Identify criteria from the rubrics or Implementation Observation Tool that contributed to your ratings.

- Describe whether the task, implementation, and teacher's questions might have provided access for all students.

- Discuss how the teacher followed up on students' responses. Identify examples of teacher's linking and teacher's press (tables 4.1, page 76, and 4.2 might be useful in this process). Identify instances where the teacher could use teacher's linking and teacher's press to increase students' contributions to the discussion.

Discuss

How do your responses compare with those in your collaborative team? What themes emerged during your discussion? In this section, we present ideas for you to consider.

Identify instances where the teacher could use teacher's linking and teacher's press to increase students' contributions to the discussion.

As you discussed your ideas with your collaborative team, perhaps you noted the interaction between the teacher's discourse actions and the nature of students' contributions. Instances of the teacher using teacher's linking or teacher's press likely increased students' opportunities to contribute to the discussion. Not only do these discourse actions provide space for students to contribute more, they also require the nature of students' contributions to go beyond answers and procedures. By consistently eliciting explanations, justifications, and connections to the mathematical work and thinking of peers, teachers create norms for what it means to participate in mathematical discussions.

Including many different students in the discussion increases students' access to the task and provides students opportunities to hear others' ways of conceptualizing the task. If you have a group of students who are unable to finish or have different conceptions about the task, you can use discourse actions to bring them into the conversation and help them to understand the thinking of their peers. Having students revoice other students' thinking can help students create their own understanding while also being active participants in the group discussion. Every student, over a few classes, should be called on to share his or her thinking and to revoice the thinking of others, so all voices are contributing to the conversation.

Hearing students' thinking also provides the teacher with opportunities to gain useful information in the formative assessment process. Making students' thinking public in the classroom community serves as a form of scaffolding. Students receive the opportunity to hear multiple perspectives and explanations and use these ideas as a way of accessing the mathematics of the lesson.

Compare the discourse actions you analyzed to the IRE discourse pattern in which teachers provide only a brief evaluation (correct or incorrect). For example, in the narratives included in the Leftover Pizza lesson version 2 (figure 2.1, page 31) and Father and Son Race lesson version 2 (figures 2.11 and 3.1, pages 50 and 58, respectively), notice how the teacher evaluates students' responses rather than using discourse actions to prompt students for more extended contributions. The teacher dominates the talking, and the teacher (rather than students) provides most of the mathematical thinking and reasoning. In both narratives, the students' roles in the discussion are to contribute correct answers and procedures and to listen to the teacher explain the mathematical ideas.

The IRE discourse pattern provides limited opportunities for students to participate in mathematical discussions and sets very different expectations for the nature of students' contributions and for students' roles in the discussion. Notice the differences in how teachers give students agency, position them as capable mathematicians, and assign them competence when the teacher's role is to evaluate students' responses and then provide the mathematics reasoning—as opposed to when the teacher uses a discourse action to elicit or extend students' ideas and contributions to the discussion. By inviting students to take up the ideas of their peers, revoicing, and focusing attention on students' thinking, teachers engage more students in the discussion. This allows students to recognize their own competence in mathematics and that of their peers.

The teacher's discourse actions in small-group and whole-class discussions also model how students should engage in talk when working in small groups when the teacher is not nearby. Encouraging students to use discourse actions helps develop the three norms featured in the *Making Sense of Mathematics for Teaching* series: (1) explain and justify solutions, (2) make sense of each other's solutions, and (3) say when you don't understand or when you don't agree. Particularly within mixed-ability grouping, students become accustomed to the expectation of asking how others are thinking, taking up the ideas of their peers, and revoicing others' thinking.

For these reasons, an awareness of teacher's linking and teacher's press provides important insight into students' opportunities to learn. In the following section, we introduce the IQA Teacher's Linking and Teacher's Press rubrics to assist in your analysis of classroom practices.

The IQA Teacher's Linking and Teacher's Press Rubrics

Linking and press support student discourse (Resnick, Asterhan, & Clarke, 2015). Just as different levels of tasks provide different opportunities for learning, different levels of linking and press provide different opportunities for students to contribute to mathematical discussions. When you use a strong linking action, you support students to take up ideas offered by their peers and generate discussion that explicitly indicates how they are making sense of those ideas. For example, an instance of strong linking occurs in the Leftover Pizza lesson version 1 (5:08):

> **Teacher:** So we see how he got seven servings; do we see how he got $\frac{1}{6}$? So do we agree with that? I'm hearing yeses and noes. . . . Who doesn't agree with him?
>
> **Student:** I disagree with him because it was six parts when it was a whole. . . .

This exchange illustrates a strong link, because the teacher prompts students to make sense of a strategy used by a peer, and we hear the student make sense of the other student's reasoning and strategy.

Revoicing alone is not considered strong linking because with this method, the teacher is sharing an idea but we do not know how other students are making sense of that idea. For example, in the Change Unknown lesson, the teacher says (1:50), "So I'm hearing you say that you knew 8 plus 7 equals 15, and you used that to know that 7 plus 8 equals 15." The teacher is providing an opportunity for students to make sense of the thinking of one of their peers, but we do not know whether they are actually doing so.

When the teacher makes a linking action and the next student's response does not directly relate back to the original speaker's idea, we consider this to be a link, but not a strong link. For example, in the Leftover Pizza lesson, the teacher asks (5:49), "Are you able to make sense of what he said?" The student responds in a way that takes up the original student's idea (that one green triangle is $\frac{1}{4}$ of a serving), but communicates his own thinking without explicitly connecting back to the original student: "One of them is $\frac{1}{4}$, since there is four parts and this is $\frac{1}{4}$" (holding up the manipulatives). To make this a strong link the teacher might have followed up with, "What did he mean by $\frac{2}{3}$?" in order to make it clear to the rest of the class that the two blue rhombus pattern blocks represented a serving of $\frac{2}{3}$ of a pizza and that the green triangle represented $\frac{1}{4}$ of that serving. Another option is for the teacher to have asked, "Can anyone else put all of those ideas together?" in order to make the connection between the students' ideas explicit.

A teacher's press generates evidence of students' thinking and reasoning. Strong press also generates evidence of students' conceptual understanding. In the Leftover Pizza lesson version 1, the teacher consistently prompts the first speaker to say more about his strategy, and the student describes his thinking and understanding of dividing $4\frac{5}{6}$ pizzas into servings of $\frac{2}{3}$ of a pizza:

> **Student:** I said that it's 7 $\frac{1}{6}$ servings. (3:12)
>
> **Teacher:** Why?
>
> **Teacher:** You made 5 circles. Okay, what did you do with them? (3:32)
>
> **Teacher:** So we have this 4 and $\frac{5}{6}$ pizzas. Then what did you do? (3:53)
>
> **Student:** I put the rest, and I split them into thirds.
>
> **Teacher:** Why?
>
> **Student:** Because, like, since each serving size is $\frac{2}{3}$, so I can actually count $\frac{2}{3}$ as one serving and another $\frac{2}{3}$ as one serving.

Following the teacher's prompts, in the video we hear evidence of the student's thinking and reasoning, specifically, of how he used the model to divide 4⅚ by ⅔. Later in the lesson, after a student refers to the "green triangle," the teacher asks the student to explain, "What did the one green triangle represent?" Understanding that the green triangle is ⅙ of a pizza but ¼ of a serving is essential to building an understanding of the meaning of remainders in fraction division, and so this action is considered an instance of strong press, intended to elicit the student's conceptual understanding. In contrast, press that elicits accurate knowledge, facts, or procedures—while often important and necessary within a mathematics lesson—is not considered strong press. Press for accurate mathematics contributes to a rating of level 2 on the Teacher's Press rubric, regardless of how many times a teacher uses such press.

The IQA Teacher's Linking rubric (figure 4.1) and Teacher's Press rubric (figure 4.2) are tools you can use for rating the specific discourse actions used during a mathematics lesson with small groups or during whole-class discussions. The rubrics provide descriptions of linking and press and assign a rating based on the number and quality of observed discourse actions. Strong discourse actions, such as those previously described in this section, are necessary for a rating of level 3 or 4.

Consistent use of strong discourse actions rates a level 4 on the Teacher's Linking and Teacher's Press rubrics, and the presence of strong discourse actions (at least twice) rates a level 3 on each of the rubrics. Teacher's press that generates accurate knowledge or procedures and teacher's linking that does not result in explicit connections between students and ideas rate a level 2. A mathematical discussion that lacks discourse actions rates a level 1.

With your collaborative team, discuss what sort of teacher discourse actions may compose strong linking or press before moving on to the application activities in the following section. In these activities we ask you to use the IQA Teacher's Linking and Teacher's Press rubrics to reflect on your current instructional practices.

IQA Teacher's Linking Rubric	
4	The teacher consistently (at least three times) explicitly connects (or provides opportunities for students to connect) speakers' contributions to each other *and* describes (or provides opportunities for students to describe) how ideas or positions shared during the discussion relate to each other.
3	At least twice during the lesson, the teacher explicitly connects (or provides opportunities for students to connect) speakers' contributions to each other *and* describes (or provides opportunities for students to describe) how ideas or positions relate to each other.
2	At one or more points during the discussion, the teacher links speakers' contributions to each other, but *does not show* how ideas or positions relate to each other (for example, implicitly building on ideas; or noting that ideas or strategies are different but not describing how).
	The teacher may revoice or recap, but *does not describe* how ideas or positions relate to each other, or makes only one strong effort to connect speakers' contributions to each other (one strong link).
1	The teacher does not make any effort to link or revoice speakers' contributions.
0	There is no class discussion, or class discussion is not related to mathematics.

Source: Adapted from Boston, 2017.

Figure 4.1: IQA Teacher's Linking rubric.

IQA Teacher's Press Rubric	
4	The teacher consistently (almost always) asks students to provide evidence for their contributions by pressing for conceptual explanations or to explain their reasoning. There are few, if any, instances of missed press, in which the teacher needed to press and did not.
3	At least twice during the lesson, the teacher asks students to provide evidence for their contributions by pressing for conceptual explanations or to explain their reasoning. The teacher sometimes presses for explanations, but there are instances of missed press.
2	Most of the press is for computational or procedural explanations or memorized knowledge, or there are one or more superficial, trivial, or formulaic efforts to ask students to provide evidence for their contributions or to explain their reasoning (for example, asking, "How did you get that?") before then moving on without attending to student responses.
1	There are no efforts to ask students to provide evidence for their contributions, and there are no efforts to ask students to explain their thinking.
0	There is no class discussion, or class discussion is not related to mathematics.

Source: Adapted from Boston, 2017.

Figure 4.2: IQA Teacher's Press rubric.

Application Activities

In the following activities, you will become familiar with the IQA Teacher's Linking and Teacher's Press rubrics as you practice rating these discourse actions.

Activity 4.4: Rating Teacher's Linking and Teacher's Press

In activity 4.4, you will identify teacher's linking and teacher's press in the Algebraic and Graphical Solutions lesson and rate these discourse actions using the Teacher's Linking and Teacher's Press rubrics.

Engage

In the Algebraic and Graphical Solutions lesson, students explore the task in figure 4.3 (page 84).

- Solve the Algebraic and Graphical Solutions to Equations task and discuss your ideas with your collaborative team before watching the lesson.
- Rate the task using the Potential of the Task rubric. Provide a rationale for your rating.
- Watch the video of the Algebraic and Graphical Solutions to Equations lesson. As you watch the lesson:
    - Note examples of teacher's linking as the teacher interacts with small groups and during the whole-class discussion
    - Note examples of teacher's press as the teacher interacts with small groups and during the whole-class discussion
- After watching the video, rate the lesson using the Teacher's Linking rubric and the Teacher's Press rubric.

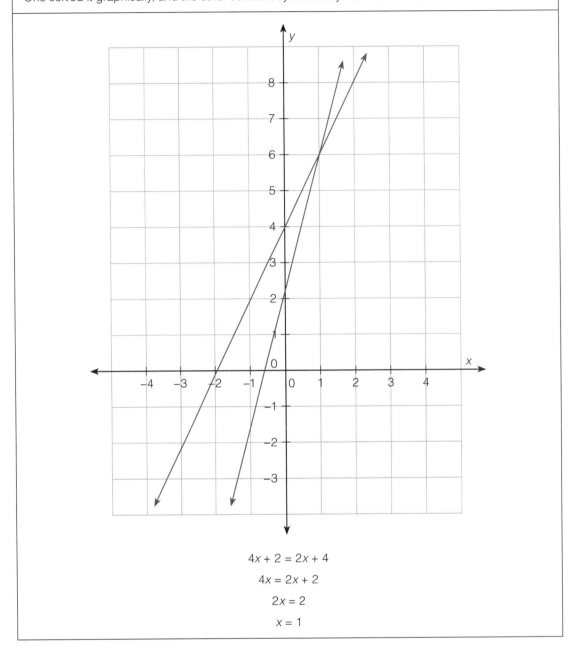

The teacher asks two students to solve for *x* given $4x + 2 = 2x + 4$.

One solved it graphically, and the other solved it symbolically. How are the solutions related?

$$4x + 2 = 2x + 4$$
$$4x = 2x + 2$$
$$2x = 2$$
$$x = 1$$

Source: Nolan, Dixon, Safi, & Haciomeroglu, 2016, p. 53.

Figure 4.3: Algebraic and Graphical Solutions to Equations task.

Discuss the examples of discourse actions you noted as well as your rubric ratings with your collaborative team before moving on to the activity 4.4 discussion.

 Algebraic and Graphical Solutions to Equations Lesson: www.SolutionTree.com/Linking_Graphical_and _Algebraic_Solutions_to_Equations

Discuss

How do your responses compare with those in your collaborative team? What themes emerged during your discussion? In this section, we present ideas for you to consider.

Rate the task using the Potential of the Task rubric. Provide a rationale for your rating.

The task rates a level 4 on the Potential of the Task rubric because students receive the opportunity to make explicit connections between representations, strategies, or mathematical concepts and procedures. By asking, "How are the solutions related?" the task explicitly prompts students to explain the connections and relationships between symbolic and graphical solutions to equations, namely, that each side of the equation can be represented by a function and graphed, and the x-value of the point of intersection of the two functions is the solution to the equation.

Note examples of teacher's linking as the teacher interacts with small groups and during the whole-class discussion. Rate the interactions using the Teacher's Linking rubric.

Teacher's linking occurs as the teacher interacts with a small group, as indicated in the following discourse:

> **Teacher:** Wait a minute, what does she mean, if you plug x into the equation you get 6? (1:11)
>
> **Student:** I don't know, what I was just doing was, I was . . .
>
> **Teacher:** So I'm not asking you what you were doing, I'm asking you to make sense of what she said.
>
> (Original student re-explains.)
>
> **Teacher:** Stop for a second. What does she mean by that?
>
> **Student:** I don't know.
>
> **Teacher:** Then you have a question to ask.

Following this exchange, the teacher uses a series of linking actions to determine if the student can use the reasoning of his peer. Because the student is then able to provide an explanation, we would consider the teacher's discourse actions in this exchange to be instances of strong teacher linking.

There are also instances of revoicing. For example:

> **Student:** So if I plug 1 in the x here and then I add them up, the solution would be 6. (2:21)
>
> **Teacher:** So that side of the equation would be 6. And what about the other side of the equation?

Note that for this activity, as an exercise in using the rubrics, we are rating teacher's linking based on the teacher's interactions with one small group and a short clip of the whole-class discussion. In the portion of the lesson portrayed, several strong links, as well as revoicing, occur within one set of interactions with students. The number and quality of teacher's linking actions would rate a level 3 on the Teacher's Linking rubric because the strong links are contained within the teacher's interaction with one small group and are not consistent throughout the lesson. If we were able to view the entire lesson and these same types of discourse actions were evident in interactions within another small group or within the whole-group discussion, Teacher's Linking would rate a level 4.

Note examples of teacher's press as the teacher interacts with small groups and during the whole-class discussion. Rate the interactions using the Teacher's Press rubric.

Teacher's press actions are consistent throughout the Algebraic and Graphical Solutions lesson. As the teacher interacts with a small group, she asks a student to explain an idea offered by a member of his group (2:10): "So what would happen when you replace the x by a 1?" In the previous interactions, the student was not able to provide this explanation. The student responds (2:12), "You get the solution for y," and the teacher then uses a series of teacher's press actions to elicit a more extended explanation from the student:

> - So then, if you plug the 1 in here, what would you get on this side of the equation?
> - So that side of the equation would be 6. And what about the other side of the equation?
> - Where is that 6 in the graph?

Through the use of teacher's press, the teacher is able to hear more of the student's thinking and better assess his understanding.

During the whole-class discussion, the teacher uses press to elicit explanations of thinking and reasoning. Consider the following exchange:

> **Teacher:** What did you find out? (5:24)
>
> **Student:** Um, if $4x = 2x + 2$ is where he stopped, the two lines would intersect at the point (1, 4).
>
> **Teacher:** What two lines are you talking about?

> ***Student:*** *The one line* y = 4x *and the other line that equals* y = 2x + 2.
>
> ***Teacher:*** *Okay, so why do you suppose they intersected at the point (1, 4)?*

In both instances, the teacher presses the student to provide additional accurate knowledge. By asking the student, "What two lines are you talking about?" the teacher provides the opportunity for the student to unpack an important idea implicit in his explanation (namely, that each side of the equation can be represented by a function and graphed as a line). This allows the teacher to determine whether the student recognizes this connection and to make this important idea explicit to the rest of the class. In the second prompt, the teacher provides an opportunity for additional explanation of the connection between the point of intersection and the solution to the equation.

The teacher also uses press to ask students to provide or clarify mathematical knowledge. For example, to prompt the student's thinking about the graph of $y = 2x$ and $y = 2$, the teacher asks (6:14), "What would the x be? . . . And what would the y be?" She then follows with press for thinking and reasoning (6:25): "Why? . . . And the point of intersection is?" In this lesson, several instances of teacher's press occur that elicit thinking and reasoning consistently throughout the small-group interactions and whole-class discussion. Hence, the lesson rates a level 4 on the Teacher's Press rubric.

How would you rate the teacher's linking and teacher's press that occurred in the lessons you observed for the chapter 3 transition activity (activity 4.3, page 79), or for the lessons featured in previous activities in this chapter: the Leftover Pizza lesson version 1 (activity 4.1, page 73) or the Change Unknown lesson (activity 4.2, page 76)? In any of these lessons, you may have noticed instances when the teacher did not follow up with linking or press. While a lack of follow-up may seem like a missed opportunity for linking or press, in the next section we consider why a teacher may decide not to follow up on a student's response.

Activity 4.5: Determining When It Is Appropriate to Ask a Follow-Up Question

During small-group interactions or whole-class discussions, a teacher might choose not to follow up on a student's response. In this activity, we will ask you to consider why a teacher may or may not choose to follow up on a student's response.

Engage

Revisit the Father and Son Race lesson version 1 (page 49). As you watch this video, do the following:

- Identify examples of when the teacher chooses *not* to press small groups early in the lesson.
- Discuss how the teacher eventually does follow up on the students' responses. Think of reasons why the teacher may have decided to follow up in this manner.

Discuss your thoughts with your collaborative team before moving on to the activity 4.5 discussion.

Discuss

Identify examples of when the teacher chooses not *to press small groups early in the lesson.*

In the Father and Son Race lesson version 1, the teacher's interactions with the first and second group differ from her interactions with the third group. With the first group, students express the common misconception in interpreting distance-time graphs that the teacher wants to elicit and address: that the son begins running faster because the line representing the son's distance and time rises above the line representing the father's distance and time (in other words, "the son starts to go faster" at point B). Because this interaction occurs early in the small-group work, she leaves the group to continue to explore the task rather than introducing new or different ideas into their initial thinking:

> **Teacher:** *So you guys agree the son starts to go faster at that point? (0:47)*
>
> **Students:** *Yes.*
>
> **Teacher:** *Okay, thanks (walks away).*

Similarly, the students in the second group (1:32) express several misconceptions in interpreting the graph, such as the father and son both starting at zero, and the father and son being "separate by 10 seconds" at C. Again, as students are still thinking through the main mathematical ideas in the task, the teacher asks questions to determine how they are thinking and purposely decides to allow them to continue to wrestle with the mathematics. While she uses linking to determine whether students in each group agree with the shared ideas, she does not press for more complete explanations at this point.

When the teacher approaches the third group (2:03), the members have had more time to consider the task. The teacher asks questions to assess their thinking and then uses press ("What are we talking about with accelerating?"; "What in the graph tells you that?") and linking ("Wait . . . then you have a question to ask"; "What does she mean by that?") to encourage students to reconsider the distance-time relationship portrayed in the graph. The teacher again chooses to leave the small group with the issue unresolved.

Discuss how the teacher eventually does follow up on the students' responses. Think of reasons why the teacher may have decided to follow up in this manner.

The teacher purposefully decides to leave the big mathematical ideas and misconceptions at the heart of the task for the whole-class discussion. She concludes the discussion with the third group by saying (3:32), "That's an interesting point. We're gonna talk about that as a whole class," and begins the whole-class discussion. In this way, the whole class of students will have the chance to wrestle with the big mathematical ideas and common misconceptions together and engage in authentic discussion, rather than only some students having the opportunity to hear a recap of their classmates' process of struggling with and making sense of the ideas.

A teacher might also choose not to follow up on students' contributions that appear to move the conversation away from the mathematical goals of the lesson. Recall the Leftover Pizza lesson version 1. In this video, the teacher chooses not to press the student who is basing his explanation on the missing piece of pizza ("I disagree with him because it was six parts when it was a whole" [5:19]). Instead, the teacher elicits an explanation that focuses on the portion of a serving (5:33): "I remember you talking about servings before. You want to talk to us about it?" In this way, the teacher provides students the opportunity to hear an explanation that can support them in making sense of the fraction division problem, rather than eliciting another explanation that reinforces the current misconception of focusing on the portion of the whole pizza.

Notice, however, that in all of the examples where the teacher decides not to follow up with linking or press, she does not revert to the IRE pattern of asking an initial question (I), hearing the student's response (R), and then providing an evaluation (E). Instead, either she allows students to continue with their discussions on their own or she returns to specific students and gauges their new understandings after other ideas are shared.

Summary

Through teacher's linking and press, you provide opportunities for students to make greater contributions to mathematical discussions in small-group or whole-class settings. By encouraging students to connect to the ideas of their peers or to provide additional detail or justification for their own explanations, teachers can support students to engage in authentic mathematical discourse. By keeping a mathematical idea in the public space of the discussion for a longer period of time, more students are able to contribute to or hear additional explanations of that idea. This increases access for each and every student and provides more students the opportunity to understand the idea more deeply.

Over time, teacher's linking and press might become less frequent as students take up these practices as norms in the classroom. As students become accustomed to being asked "Why?" or prompted to "Say more," they may begin to offer extended explanations of their thinking and reasoning during their initial contributions and may not need to be pressed for additional detail or explanation. For example, the first student to contribute to the whole-group discussion in the Algebraic and Graphical Solutions lesson (4:12 to 4:36) provides a thorough explanation of the importance of the point of intersection. In cases like these, where extra press may be unnecessary, the teacher's next action is to ask a new question that will provide students an additional opportunity to consider the importance of the point of intersection: "What would have happened if the student started to solve it symbolically but then graphed at $4x = 2x + 2$? What would the point of intersection be at that point?" Similarly, with teacher's linking, when students anticipate that you will ask them to connect to, make sense of, or restate another student's idea, they begin to provide these connections without prompting.

When you use linking and press, you are able to extend the time and quality of students' contributions to discussions, to elicit deeper explanations or clear and accurate knowledge, to make implicit ideas explicit in the conversation, and to promote student-to-student interactions. As you transition to chapter 5, in addition to your discourse actions, you will also consider the nature of students' contributions and the evidence of thinking and reasoning provided by students as they work in small groups and engage in whole-class discussions.

Chapter 4 Transition Activity: Teacher's Linking, Teacher's Press, and Students' Contributions

Work within your collaborative team to observe or video record each of you teaching a mathematics lesson. You may want to pair up and observe each other or share videos. Obtain a copy of the task and take notes during the observation or as you view the video. For each lesson, do the following.

- Rate the Potential of the Task and the Implementation of the Task using the rubrics. How do the task and implementation provide access for all students?

- Write down examples of linking and press that the teacher used to follow up on students' contributions. Write down students' responses as well. Try to capture the exact wording used by the teacher and students. This will be important in chapter 5 when we ask you to consider the nature of students' contributions.

- In what ways did the teacher follow up on students' contributions? Did the teacher use linking to prompt students to build on the ideas of their peers? Did the teacher use press to prompt students to provide more extended explanations or justifications?

- Following linking or press by the teacher, describe whether students actually built on or related to the ideas of their peers or provided explanations and justifications.

Save your responses, ratings, and notes or any written reflections from the transition activity, as you will refer to them in chapter 5.

PART 3
Connecting to the *E* in TQE: Evidence of Students' Mathematical Work and Thinking

In part 1, you analyzed *tasks* and their implementation using the Potential of the Task rubric and the Implementation of the Task rubric. In part 2, you analyzed instructional quality by exploring teacher's *questions* and other discourse actions. Part 3 connects to the *E* in the TQE process: "Evidence: Collect and use evidence of student understanding in the formative assessment process to guide the delivery of instruction" (Dixon, Nolan, & Adams, 2016, p. 4). In part 3, you will examine evidence or outcomes of the teacher's discourse actions in the form of students' verbal contributions within small-group and whole-class discussions. You will consider the impact of teachers' questions (discussed in chapter 3) and follow-up (addressed in chapter 4) on students' opportunities to engage with mathematics and contribute to mathematical discussions.

Both rubrics introduced in chapter 4 (Teacher's Linking and Teacher's Press) have companion IQA rubrics that you will explore in chapter 5 as evidence of students' engagement with the mathematics. The Students' Linking rubric provides a tool to analyze students' connections to the work and thinking of their peers, and the Students' Providing rubric provides a tool to analyze students' mathematical justifications and explanations.

In chapter 5, you will identify evidence of students' mathematical work and thinking apparent in students' representations, strategies, explanations, and justifications. In this chapter, you will explore evidence in the form of students' verbal contributions during discussions. The contributions take the form of students' linking and students' providing. Both types of evidence are student outcomes during instruction.

In chapter 6, we will ask you to use all the rubrics presented throughout the book and collect evidence to pull it all together. In this chapter, you and your collaborative team will explore evidence from classrooms, including your own, and apply the IQA rubrics to reflect on instructional practice.

CHAPTER 5
Students' Linking and Students' Providing

The way teachers and students talk with one another in the classroom is critical to what students learn about mathematics and how they come to see themselves as mathematical thinkers.

—Elham Kazemi and Allison Hintz

In this chapter, you will explore how to analyze the level and quality of students' contributions to mathematical discussions by examining students' explanations and justifications and whether students make connections to the mathematical work and thinking of their peers. You will focus on the responses students provide—to tasks or discourse actions—as evidence of students' thinking and reasoning. You will be able to answer the following questions.

- What types of contributions do students make to small-group or whole-class discussions?
- What evidence of students' thinking and reasoning is evident during the small-group or whole-class discussions?
- In what ways do students explicitly build on the work and ideas of their classmates?

Introductory Activities

Let's get started by thinking about different ways students might show evidence of their learning in their responses to tasks and teacher's questions. In activities 5.1 through 5.3, we ask you and your collaborative team to consider several sample lessons and to analyze how students respond and what the teacher may learn from their responses.

Activity 5.1: Examining Students' Contributions—Students' Linking

In activity 5.1, we ask you to think about how students contribute to solving a problem in a pulled small group and what this outcome tells you about instructional quality.

Engage

Activity 5.1 involves a third-grade classroom exploring elapsed time in a variety of centers, and one of the centers is a pulled small group (Dixon, Brooks, & Carli, 2018). The Elapsed Time lesson features the teacher and four students engaging with the following task: "Sasha is taking a flight from Orlando, Florida, to Manchester, New Hampshire. The plane is scheduled to take off at 7:15 a.m. and land at 11:45 a.m. How long should it take Sasha to fly from Orlando to Manchester?" (Dixon, Nolan, Adams, Tobias, & Barmoha, 2016, p. 150).

Complete the task and consider the following prompts.

- Rate the task using the Potential of the Task rubric.
- Watch the Elapsed Time lesson with a focus on students' contributions. Write down examples of students' contributions in which students relate to or build on each other's ideas.
- Rate the overall lesson using the Implementation of the Task rubric.

Discuss your ratings and examples with your collaborative team before moving on to the activity 5.1 discussion.

Elapsed Time Lesson:
www.SolutionTree.com/GR3ElapsedTime

Discuss

How do your responses compare with those in your collaborative team? What themes emerged during your discussion? In this section, we present ideas for you to consider.

Rate the task using the Potential of the Task rubric. Rate the overall lesson using the Implementation of the Task rubric.

The Potential of the Task rates a level 3 as there are many ways students could solve the task, including using number lines, a clock, or student-invented strategies. Students do not receive a set procedure for how to solve the task, and there is an element of problem solving inherent in the task for third-grade students. The task does not rate a level 4 because it does not explicitly prompt for students' thinking and reasoning.

The Implementation of the Task rates a level 4. As with previous tasks, even though the task does not explicitly prompt for students' thinking and reasoning, the teacher uses discourse actions that make students' thinking visible during the lesson as students provide explanations of their strategies.

In what ways do students relate to and build on each other's ideas?

At several points during the teacher's interaction with the small group featured in the Elapsed Time lesson, students explain ideas originally offered by one of their peers. For example, following the teacher's prompt, Zaineb explains how another student used the number line to solve the problem:

> **Teacher:** Why do you think he added 5 and then added 40? (1:48)
>
> **Zaineb:** He added 40 minutes to get to the next hour.
>
> **Teacher:** Why would he want to do that?
>
> **Zaineb:** So it's easier for him to keep track.

Zaineb originally described her own strategy of making jumps of five minutes and needing a larger number line. The teacher then created an opportunity for her to make sense of a different strategy. At timestamp 2:04, the teacher again follows up with, "So, who would have to make more jumps?" Zaineb replies that she would need to make more jumps, because "I'm adding by 5 and he's adding by 40." Later in the discussion (3:15), the teacher creates similar opportunities, and students explicitly connect to the ideas of their peers:

> **Teacher:** *So you've all done different things here. . . . Solomon did this* (referring to the diagram created by Solomon on the whiteboard), *but Santiago, can you tell us what he's done?*
>
> **Santiago:** *So, he started from 7:15, and then he added 5 minutes and he got 7:20. And then from 7:20, he added 40 minutes to get to 8 o'clock.*
>
> **Teacher:** *Hold on. Zaineb, can you conjecture why you think he added 40 minutes here?*
>
> **Zaineb:** *He, so you could, like, take less time of getting to 8 o'clock instead of adding 5 and 5.*

The students' contributions in this exchange are examples of students' linking because we explicitly hear students explaining ideas offered by their peers. In each instance, the teacher created the opportunity for the student to make the connections (teacher's linking). However, the students themselves can also initiate instances of students' linking without prompting from the teacher. By hearing students explain, relate to, or build on each other's ideas, the teacher gathers evidence of students' mathematical thinking and understanding. In the TQE process, the teacher then incorporates this evidence of students' mathematical thinking and understanding into a formative assessment cycle.

Teachers also gain evidence of students' mathematical thinking and understanding by prompting them for explanations and justifications. In chapter 4, you identified teacher's press that elicited deeper explanations from students. In activity 5.2, the focus switches from discourse actions that the teacher makes to the explanations and justifications the students provide.

Activity 5.2: Examining Students' Contributions—Students' Providing

In this activity, you will have the opportunity to examine grade 3 students' contributions in the Squares and Rectangles lesson.

Engage

Watch the video of the Squares and Rectangles lesson. This lesson features a task in which the teacher asks the students to create a rectangle and then a square with manipulatives. As you watch the lesson, consider the following.

- In what ways do students provide explanations and justifications of their mathematical work and thinking?

- Write down examples of students' explanations and justifications.

Squares and Rectangles Lesson: www.SolutionTree.com/Defining_and_Classifying_Squares_and_Rectangles

Discuss your examples with your collaborative team before moving on to the activity 5.2 discussion.

Discuss

How do your responses compare with those in your collaborative team? What themes emerged during your discussion? In this section, we present ideas for you to consider.

In what ways do students provide explanations and justifications of their mathematical work and thinking? Write down examples of students' explanations and justifications.

There are many instances in which the students provide explanations and justifications of their work related to squares and rectangles. In the Squares and Rectangles lesson, after the teacher asks the class to think more about whether a square is a rectangle, a student explains why a square is a rectangle (4:01):

> *You know, this is where I am and my group is because, you know, a square has four right angles and a rectangle has four right angles. And a square has opposite sides are equal because if a square has these sides are equal, they're opposite and they're equal. And these sides are equal because the opposite is equal.*

Later in the lesson, the teacher challenges students to "make a square that is not a rectangle." When the class comes back together, a student explains (5:20):

> *I think that since the purple and now I think it can be a rectangle now. Because they all have four right angles, the opposite sides are equal, and they both, like, it doesn't matter how long they are; it's just the opposite sides have to be equal. And they all have four right angles.*

Throughout the discussion, students provide explanations about why squares are always rectangles, but some rectangles are not squares. Students also offer evidence for their claims about the relationship between a square and a rectangle.

As you watched the lesson, did you also notice teacher's linking and students' linking? An example of teacher's and students' linking occurs when the teacher first calls the class back together and asks whether they have all created squares (0:44). A student responds, "I think hers isn't [a square] because her red sides aren't equal to the blue sides." Later in the lesson, the teacher makes a linking action to focus the class on a student's explanation by asking (4:42), "Now what did he say?" A student responds by linking to her classmate's work and idea and also justifying her own ideas: "He said that a rectangle can be a square sometimes. But a rectangle that he has is not a square because only squares can be rectangles. Rectangles only sometimes can be squares."

Next, you will have the opportunity to consider students' linking and students' providing instances that occurred in the lessons you observed or video recorded for the chapter 4 transition activity.

Activity 5.3: Revisiting the Chapter 4 Transition Activity—Teacher's Linking, Teacher's Press, and Students' Contributions

In the chapter 4 transition activity (page 90), you analyzed the task, implementation, and discourse actions that occurred during a mathematics lesson. In the next activity, we ask you to discuss your ideas with your collaborative team.

Engage

With your collaborative team, do the following.

- Discuss your rating for Potential of the Task, Implementation of the Task, and Teacher's Questions. Identify criteria from the rubrics or Implementation Observation Tool that contributed to your rating.

- Describe whether the task, implementation, and teacher's questions provided access for all students.

- Discuss how the teacher followed up on students' contributions. Identify examples of teacher's linking and teacher's press. Identify instances where teacher's linking and teacher's press could have served to increase students' contributions to the discussion and discuss what discourse actions the teacher could have used.

- Discuss whether students actually built on or related to the ideas of their peers or provided explanations and justifications. Identify examples of students' linking and students' providing.

Discuss

As noted in chapter 4, perhaps you and your collaborative team identified connections between teacher's discourse actions and students' opportunities to contribute to the discussion. By eliciting students' contributions, teachers are able to hear evidence of students' thinking and reasoning. When students provide explanations of their own thinking or make connections to the thinking of their peers, teachers can hear whether and how students are making sense of the main mathematical ideas in the lesson.

Sometimes, eliciting evidence through connections and explanations may require multiple discourse actions on the part of the teacher, especially as students are becoming accustomed to linking to the ideas of their peers and providing explanations and justifications of their own mathematical thinking and reasoning. Often in the video clips, the teacher provides an opportunity for students to link to the ideas of others, but the students instead want to explain their own thinking. In these instances, such as in the example from the Algebraic and Graphical Solutions lesson (page 85), we hear the teacher follow up with a second instance of teacher's linking to help the students focus on linking:

> **Teacher:** *Wait a minute, what does she mean, if you plug x into the equation you get 6? (1:11)*
>
> **Student:** *I don't know, what I was just doing was, I was . . .*

> **Teacher:** So I'm not asking you what you were doing, I'm asking you to make sense of what she said.

Similarly, a teacher's press can elicit different types of explanations from students. A prompt such as "How did you know the answer is 54?" could elicit an explanation that provides evidence of knowledge of facts and procedures, such as "I know 9 times 6 is 54," or an explanation that provides evidence of conceptual understanding, such as "There are 9 packages of yogurt with 6 cups of yogurt in each package." When students respond with facts and procedures, the teacher might press again to determine if the student understands *why* the fact or procedure makes sense: "Why does it make sense to multiply in this problem?" or "How did you know to multiply 9 times 6?"

A teacher's linking and press creates space in the discussion for students' linking and providing. Consider the different versions of the Father and Son Race lesson in chapters 2 and 3. In version 2 of the lesson provided in figure 3.1 (page 58), rather than following up on a student's response with linking or press, the teacher often evaluates the correctness of the response, provides an explanation to students, and moves on to the next question:

> **Teacher:** Okay, class, let's talk about your stories for the Father and Son Race. Who started here at zero (pointing to the graph)? Juan?
>
> **Juan:** The son.
>
> **Teacher:** That's right, the son started at zero. And who started at 20? Desiree?
>
> **Desiree:** The father.
>
> **Teacher:** Correct. So at the start of the race, the son gave his father a head start. Then they start racing and meet here (points to the point of intersection). What do we call this point? Chen?
>
> **Chen:** The point of intersection.
>
> **Teacher:** What do we know is the same at this point? Alex?
>
> **Alex:** They ran the same distance in the same time.
>
> **Teacher:** Excellent answer! At the point of intersection, they are at the same distance and time. What happens after this point? Sam?
>
> **Sam:** The son starts running faster and takes the lead.
>
> **Teacher:** Well, not quite. Remember at the beginning of the lesson, we said that the father and son are running at constant speeds. . . .

Students' contributions are brief, and we do not hear evidence of students' thinking and reasoning. In fact, the teacher explains the main mathematical ideas. In contrast, consider the following exchange from the Father and Son Race lesson version 1 (3:45):

> **Student:** They both started at the time 0, but at the meters, the dad was ahead 20 meters.
>
> **Teacher:** What on the graph told you that?
>
> **Student:** The fact that it started higher than the son's (pointing to the graph), when you point here to the 0 (indicates the y-axis).

In this exchange, the teacher's press provides an opportunity for the student to justify his connection between the graph and the context. In doing so, other students benefit by hearing the explanation, the student providing the explanation benefits by verbalizing his ideas, and the teacher receives evidence of the student's understanding.

Across grade levels, students' initial responses are often brief. Eliciting connections and explanations can require multiple discourse actions from the teacher. Consider the following exchange from the Father and Son Race lesson version 1 (page 49):

> **Teacher:** What is happening with the speed? (4:35)
>
> **Student:** They're increasing.
>
> **Teacher:** What is increasing?
>
> **Student:** The father and the son.
>
> **Teacher:** How do you know?
>
> **Student:** 'Cause the graphs of the lines are going up.
>
> **Teacher:** And that indicates that they are increasing in speed?
>
> **Student:** Mm-hm [yes].
>
> **Teacher:** Agree or disagree?

In this exchange, the student provides brief responses, and the teacher continues to press until it is clear that the student holds a misconception shared by many of her classmates (namely, that the speed is increasing because the lines are "going up"). The teacher then makes a linking action by saying, "Agree or disagree?" so that other students can express their thinking about this idea. As more students have the opportunity to provide explanations or hear the explanations of others, more students are likely to develop an understanding of the main mathematical ideas.

Over time, as connecting to the ideas of others and providing thorough explanations become the norm and expectation of the class, teachers may find that fewer or less frequent discourse actions are necessary as students initiate their own links to the ideas of their classmates and provide their reasoning in their initial response along with their answers. For example, in the Father and Son Race lesson version 1, the teacher asks (4:13), "What is happening at B on this graph?" The student provides a thorough initial response: "That is the intersection, where the father and son are at the same distance and the same time" (pointing to the graph). The teacher then determines that she does not need to press further.

Think about how you might provide feedback to the teachers in these exchanges if they were peers in your collaborative team. In general, when providing feedback on teachers' linking and press actions to those in your collaborative team, tables 4.1 (page 76) and 4.2 (page 78) can offer examples. In the following section, we present two additional rubrics to help you assess student responses: the IQA Students' Linking and Students' Providing rubrics.

The IQA Students' Linking and Students' Providing Rubrics

The IQA Students' Linking rubric (figure 5.1) and Students' Providing rubric (figure 5.2) are used to rate the level of students' contributions during small-group interactions or during whole-class discussion. The rater identifies students' contributions that occur during small-group or whole-class discussions and assigns a rating based on the number and quality of observed contributions that count as students' linking or students' providing.

IQA Students' Linking Rubric	
4	The students consistently, explicitly connect their contributions to each other and describe how ideas or positions shared during the discussion relate to each other (for example, "I agree with Sam because . . .").
3	At least twice during the lesson, students explicitly connect their contributions to each other and describe how ideas or positions shared during the discussion relate to each other (for example, "I agree with Mohammed because . . .").
2	At one or more points during the discussion, the students link students' contributions to each other, but do not describe how ideas or positions relate to each other (for example, implicitly using or building on others' ideas; or "I disagree with Ana"), or students make only one strong effort to connect their contributions with each other.
1	Students do not make any effort to link or revoice students' contributions.
0	Class discussion is not related to mathematics, or there is no class discussion.

Source: Adapted from Boston, 2017.

Figure 5.1: IQA Students' Linking rubric.

IQA Students' Providing Rubric	
4	Students consistently provide evidence for their claims, or explain their thinking using reasoning in ways appropriate to the discipline (for example, conceptual explanations).
3	Once or twice during the lesson, students provide evidence for their claims, or explain their thinking using reasoning in ways appropriate to the discipline (for example, conceptual explanations).
2	Students provide explanations that are computational, procedural, or memorized knowledge, or what little evidence or reasoning students provide is inaccurate, incomplete, or vague.
1	Students do not back up their claims or do not explain the reasoning behind their claims.
0	Class discussion is not related to mathematics, or there is no class discussion.

Source: Adapted from Boston, 2017.

Figure 5.2: IQA Students' Providing rubric.

Recall that when rating teacher's linking and teacher's press, strong discourse actions are necessary for a rating of level 3 or 4. In other words, a rating of level 3 or 4 requires teacher's linking actions that explicitly indicate how students are making sense of the ideas of their peers. A teacher's press that elicits explanations of students' conceptual understanding is necessary for a rating of level 3 or 4.

The Students' Linking and Students' Providing rubrics are rated in similar ways. Explicit *links* to other students' ideas (such as the use of names or pronouns to identify another student's ideas or strategy, or explicit explanations of or comparisons to another student's work) are necessary to rate a level 3 or 4 on the Students' Linking rubric. Explanations and justifications that *provide evidence* of conceptual understanding are necessary to rate a level 3 or 4 on the Students' Providing rubric.

In both rubrics, the consistent occurrence of strong student contributions rates a level 4 and the presence of strong student contributions (at least twice) rates a level 3. Student explanations and justifications that provide evidence of accurate knowledge or procedures and students' linking that only implicitly connects with a peer's idea rate a level 2. A discussion with a lack of students' linking rates a level 1 on the Students' Linking rubric, and a discussion with a lack of students' explanations and reasoning rates a level 1 on the Students' Providing rubric.

Using the rubrics, rate the lessons you observed for the chapter 4 transition activity (page 90). Discuss your ratings and examples with your collaborative team before proceeding to the application activity. In activity 5.4, you will have the opportunity to practice using the rubrics to rate a new lesson: the Decimals on a Number Line lesson.

Application Activities

In the following activity, we ask you to consider the nature and quality of students' responses using the Students' Linking and Students' Providing rubrics.

Activity 5.4: Rating Students' Linking and Students' Providing

In this activity, you will use the Students' Linking and Students' Providing rubrics to rate students' responses.

Engage

In activity 5.4, you will observe fourth-grade students exploring the location of decimals to hundredths on a number line. Students are working in a variety of centers, and one of the centers is a small group (Dixon, Brooks, & Carli, 2018). The lesson features a small group of four students and the teacher engaging with the Decimals on a Number Line task:

> Problem 1: Draw a number line between zero and one and estimate the location of 0.7.
>
> Problem 2: Locate and label a decimal between 1.7 and 1.9.
>
> Problem 3: Locate and label a decimal between 0.3 and 0.4 on the number line. (Dixon, Brooks, & Carli, 2018, p. 32)

With your collaborative team, do the following.

- Solve the Decimals on a Number Line task.
- Rate the task using the Potential of the Task rubric. Provide a rationale for your rating.
- Watch the video of the Decimals on a Number Line lesson. As you watch the lesson, note examples of students' linking (to the ideas of their peers) and students' providing (explanations and justifications) during small-group discussion.
- After watching the lesson, use the Students' Linking and Students' Providing rubrics to rate the lesson.

Discuss your examples and rubric ratings with your collaborative team before moving on to the activity 5.4 discussion.

Decimals on a Number Line Lesson:
www.SolutionTree.com/GR4DecimalHundredths

Discuss

How do your responses compare with those in your collaborative team? What themes emerged during your discussion? In this section, we present ideas for you to consider.

Rate the task using the Potential of the Task rubric. Provide a rationale for your rating.

The Potential of the Task is rated a level 3 because the task has the potential to engage students in complex thinking about concepts related to decimals and the relationships between tenths and hundredths. Specifically, students are asked to use the number line to make sense of the relationship between tenths and hundredths and how tenths and hundredths are represented on a number line. The task is not rated a level 4 because there is not an explicit prompt for students' reasoning of why or how they located and labeled their decimals.

Note examples of students' linking during small-group discussion. Use the Students' Linking rubric to rate the lesson.

There are several examples of strong students' linking in the Decimals on a Number Line lesson. Two strong instances of linking occur in the following exchange, in the responses from Joseph and Abby:

> **Teacher:** So now that each of you has had the opportunity to finish, let's look at each of yours and talk about what's the same and what's different. Joseph? (1:52)
>
> **Joseph:** What's different from mine and Lina's is that she didn't write the 0 to 2; she did the one and seven-tenths and one and nine-tenths and it would be half for one and eight-tenths.

> **Teacher:** Hmm, what did he say?
>
> **Abby:** He said for part of the number line she took part of it, so if I erased all of this, this is what she would have.

Instances of students' linking also occur when Cesar makes a link to Abby's work:

> **Cesar:** Um, she put zero point thirty-five-hundredths because three-tenths is equal to thirty-hundredths. (4:08)
>
> **Teacher:** Stop for a moment. What did he say? Is he correct?
>
> **Abby:** I don't know. I'm still stuck. I don't understand. Both of us are kind of at the same point. We both understand, we are saying the same thing, it's just like we don't know if it's a tenth or a hundredth.

Later in the lesson, Joseph also makes a link to Abby's work (4:52): "Yes, she is actually correct because zero and three-tenths would be thirty." With strong student links consistent throughout the lesson, this small-group discussion would rate a level 4 on the Students' Linking rubric. Students connect their contributions to each other and describe how they are making sense of their peers' ideas.

Note examples of students providing explanations and justifications during small-group discussion. Use the Students' Providing rubric to rate the lesson.

There are many instances of students providing explanations of how they determined the placement of decimals on the number line. In the first part of the task, in which students are asked to place seven-tenths on the number line, the following teacher's press and students' providing occurs:

> **Teacher:** So I am going to pull this out so we can talk about it. I see that you wrote zero and one and you started to make the seventh hash tag here. (0:49)
>
> **Joseph:** But they were very small (he has seven very small hash marks on his number line).
>
> **Teacher:** Okay, what do you mean by that?
>
> **Joseph:** 'Cause that's very small and the one is all the way over here and the seven-tenths should be right here (points to a place closer to one).

Other students also provide justifications for the location of seven-tenths on the number line:

> **Cesar:** Because the seven-tenths should be closer to one than closer to zero. (1:11)
>
> **Lina:** Because seven-tenths is greater than five-tenths. (1:18)

Later in the lesson, Joseph also justifies why Abby is correct that thirty-five-hundredths would be between three-tenths and four-tenths (4:52):

> *Because zero and three-tenths would be thirty . . . thirty-hundredths and then in the middle of zero and three-tenths and zero and four-tenths would be thirty-five-hundredths because you are telling us to label one between zero and three-tenths and zero and four-tenths.*

Because students consistently provide evidence for their claims and explain their thinking, the lesson rates a level 4 on the Students' Providing rubric.

As you watched the lesson, you may have noted how teacher's linking and press support students' linking and providing. You may also have noted instances when students connect to another student's ideas or provide an explanation without being prompted by the teacher. When the teacher notices Joseph's desire to contribute to the discussion, Joseph begins his explanation at timestamp 4:52 by saying, "Yes, she is actually correct because . . ." thus connecting to the idea of his classmate and providing a justification on his own.

As you watched the lesson, did you also consider how other rubrics impact the nature of students' contributions to the discussion? For example, tasks at a level 3 or 4 on the Potential of the Task rubric provide something for students to think, reason, and talk about. Tasks at a level 1 or 2, with a focus on procedures and memorization, provide few opportunities for mathematical connections, explanations, or justifications beyond discussing whether an answer or procedure is correct or incorrect. Similarly, a lesson rated a level 1 or 2 on the Implementation of the Task rubric or Teacher's Questions rubric would not have provided many opportunities for students to share or express their mathematical work and thinking, thus limiting students' providing of explanations and justifications and the opportunity for other students to make connections to those explanations and justifications. Implementing a task and asking questions that encourage thinking and reasoning create space for students' contributions—and evidence of students' mathematical understanding—throughout a lesson.

Summary

The rubrics in this chapter are different in nature from the rubrics in the other chapters due to their focus on students' contributions. The selection of tasks, implementation of tasks, and discourse actions are direct instructional practices of the teacher. We can consider students' linking and students' providing the *result* of the opportunities teachers give students to engage in sense making and understanding. Students' linking and students' providing offer a window into students' thinking. Reflecting on students' contributions during a lesson provides evidence of students' learning. You can use evidence of students' learning to plan upcoming lessons as well as to reflect on your practice.

Chapter 5 Transition Activity: Using All IQA Rubrics

Work within your collaborative team to video record each of you teaching a mathematics lesson. Share the videos and rate each other's lessons using the following directions. You can choose to watch and rate the videos as a group or in pairs.

- Before watching each video, obtain a copy of the task, and rate the task with the Potential of the Task rubric.
- As you view each video, take notes on instructional practices, discourse actions, and students' responses that will provide evidence for rating the video using the rubrics. You may want to watch the video once without stopping, then rewatch the video and pause to take notes.
  - Write down examples of questions that seemed to promote students' opportunities for thinking and reasoning and questions that seemed to elicit procedures, facts, or short responses.
  - Write down instances of linking and press used by the teacher to follow up on students' contributions.
  - Write down examples of students' linking to the ideas of their peers and students' providing explanations and justifications.
- After viewing each video, rate the lesson using the Implementation of the Task rubric, and use the Implementation Observation Tool to support your rating. How do the task and implementation provide access for all students? Rate the Teacher's Questions, Teacher's Linking, Teacher's Press, Students' Linking, and Students' Providing using the associated rubrics and based on the examples of the discourse actions you identified.
- Reflect on the following questions.
  - In what ways did the teacher follow up on students' contributions? Did the teacher use linking actions to prompt students to build on the ideas of their peers? Did the teacher use press to prompt students to provide more extended explanations or justifications? Provide feedback regarding any missed opportunities.
  - Following a linking or press by the teacher, describe whether students actually built on or related to the ideas of their peers, or provided explanations and justifications.

Save your responses, ratings, and notes or any written reflections from the transition activity, as you will refer to them in chapter 6.

CHAPTER 6

The IQA Toolkit as a Tool to Assess and Improve Instructional Practice

Focused reflection on your daily practice has the potential to improve mathematics instruction and lead to more high-quality, equitable mathematics instruction for your students.

—Eileen G. Merritt, Sara E. Rimm-Kaufman, Robert Q. Berry III,
Temple A. Walkowiak, and Erin R. McCracken

Throughout this book, you have considered the nature of the TQE process through a focus on tasks and implementation of tasks, teacher's questions and discourse actions, and evidence through students' responses. In this chapter, all of the rubrics from the previous chapters come together as a toolkit that will help you reflect on and identify pathways for improving instruction.

In this chapter, you will explore how to use the Instructional Quality Assessment (IQA) rubrics to reflect on instructional practice over time. You will be able to use the IQA as a tool to:

- Assess and reflect on mathematics instructional practice
- Provide feedback within your collaborative team around a common set of instructional practices using shared language and understandings
- Identify pathways for instructional improvement
- Collect data on instructional practices over time
- Monitor instructional efforts to reach each and every student

Introductory Activities

Let's get started by using the IQA rubrics to assess instruction in a pulled small-group setting (activity 6.1) and a whole-class lesson (activity 6.2).

Activity 6.1: Rating a Small-Group Lesson

In this activity, we ask you to revisit the Decimals on a Number Line lesson using the Teacher's Questions, Teacher's Linking, and Teacher's Press rubrics.

Engage

In chapter 5, you considered students' responses in the Decimals on a Number Line lesson (page 102). Recall that the task rated a level 3 on the Potential of the Task rubric because students can use the number line to make sense of the relationship between tenths and hundredths but there is not an explicit prompt for students' reasoning (as required for level 4). In the activity 5.4 discussion, we also rated the lesson as a level 4 for Students' Linking and Students' Providing.

- Rewatch the Decimals on a Number Line lesson (page 102). As you watch the lesson, identify examples of discourse actions. Write down what the teacher says and record the timestamps that provide instances of teacher's questioning, teacher's linking, or teacher's press. Because these rubrics are based on a count of discourse actions, it is important to identify specific examples.
- After watching the lesson, rate it using the Implementation of the Task rubric, and use the Implementation Observation Tool to support your rating. Rate the lesson using the Teacher's Questions, Teacher's Linking, and Teacher's Press rubrics, based on the examples of discourse actions you identified.
- Provide a rationale for your rating of each rubric. You can use the IQA summary sheet in figure 6.1 as a template.
- Use the IQA rubrics to identify feedback you would provide to the teacher.

Discuss your ratings and rationales with your collaborative team before moving on to the activity 6.1 discussion.

Dimension	Rating	Rationale and Instances
Potential of the Task		
Implementation of the Task		
Teacher's Questions		
Teacher's Linking		
Students' Linking		
Teacher's Press		
Students' Providing		

Figure 6.1: IQA summary sheet.

Visit go.SolutionTree.com/mathematics *for a free reproducible version of this figure.*

Discuss

How do your responses compare with those in your collaborative team? What themes emerged during your discussion? In this section, we present ideas for you to consider.

Rate the lesson using the Implementation of the Task, Teacher's Questions, Teacher's Linking, and Teacher's Press rubrics. Provide a rationale for the rating of each rubric.

We provide ratings and rationales for the Potential of the Task, Students' Linking, and Students' Providing rubrics in the activity 5.4 discussion (page 102). Here we discuss the Implementation of the Task, Teacher's Questions, Teacher's Linking, and Teacher's Press rubrics.

IMPLEMENTATION OF THE TASK

Even though the task does not prompt students to explain their reasoning, the teacher uses several discourse actions that make students' thinking and reasoning explicit during the small-group discussion. Recall that Implementation of the Task is an overall holistic rating of students' level of engagement in cognitively demanding mathematical work and thinking throughout the lesson. From the Implementation of the Task rubric, students *solve a genuine, challenging problem for which students' reasoning is evident in their work on the task*, as they are challenged by and make sense of decimals in the hundredths between 0.3 and 0.4. Students, rather than the teacher, make explicit connections between the number line (representation) and decimals in the tenths and hundredths (the mathematical concept of place value). While students all use a number line, students *follow a prescribed procedure in order to explain or illustrate a mathematical concept*. Hence, the lesson rates a level 4 on the Implementation of the Task rubric.

While many points in section A on the Implementation Observation Tool characterize the lesson, these two seem particularly salient.

1. *The teacher provides consistent requests for explanation and meaning*, as identified in the examples provided for teacher's press.

2. *The teacher provides students with sufficient modeling of high-level performance on the task* through the use of number lines created by students and their peers.

During the discussions, we also hear evidence of the following items from section C: *making explicit connections or comparisons between strategies, or explaining why they chose one strategy over another*, and *making connections between a representation* (number lines) *and the underlying mathematical ideas* (place value representations to the tenths and hundredths).

TEACHER'S QUESTIONS

During the lesson, the teacher asks academically relevant questions that provide opportunities for students to elaborate and explain their mathematical work and thinking (*probing students' thinking* and *generating discussion*), identify and describe the important mathematical ideas in the lesson, or make connections between ideas, representations, or strategies (*exploring mathematical meanings and relationships*). Every time the teacher pulls the students back together for a discussion, she asks an initial question to elicit students' thinking and reasoning:

- *So, I am going to pull this out so we can talk about it. I see that you wrote 0 and 1 and you started to make the seventh hash tag here* (student begins talking). *(0:49)*

- *Let's look at each of yours and talk about what's the same and what's different. Joseph? (1:52)*

- *So, Lina, I see that you have made a mark between three-tenths and four-tenths and now you've paused. What are you thinking? (3:00)*

In many instances, the teacher's questions also serve the purpose of linking or press. For example, the initial question to begin the discussion at 1:52 is also an instance of teacher's linking. Note that the "initial question" at 0:49 is not actually a question, as the student begins to talk before the teacher finishes her prompt.

Throughout the lesson, following the initial questions to bring the group back together (at 0:49, 1:52, and 3:00), most of the teacher's discourse actions are intended to press students to make sense of decimals, to explain their number lines and ideas, or to utilize linking to make sense of each other's work and thinking. Hence, in the lesson we hear mostly linking and press rather than the consistent use of new questions. Thus, the lesson would rate a level 3 on the Teacher's Questions rubric. Interestingly, according to classroom research, teachers use linking and press far less frequently than new or initial questions (Boston & Wilhelm, 2015). More often, they plan and ask a variety of questions—often very good questions—but do not follow up on students' responses with linking or press.

When using the IQA rubrics, first identify examples of different types of questions the teacher is asking, particularly questions that align with the questioning types of *probing students' thinking, generating discussion*, and *exploring mathematical meanings and relationships*. Then, determine if questions are serving the purpose of linking or press, and decide how to rate the balance between the questions and linking and press. Typically, lessons that rate highly (level 3 or 4) for Teacher's Linking and Teacher's Press will also rate highly (level 3 or 4) for Teacher's Questions. However, lessons that rate highly for Teacher's Questions may or may not rate highly for Teacher's Linking or Teacher's Press, because those rubrics look for very specific types of discourse actions.

Teacher's Linking

The following actions count as instances of strong teacher linking, because in each case the student's response that follows makes an explicit connection to another student's work or ideas:

> **Teacher:** Let's look at each of yours and talk about what's the same and what's different. Joseph? (1:52)
>
> **Joseph:** What's different from mine and Lina's is that she didn't write the 0 to 2; she did the one and seven-tenths and one and nine-tenths and it would be half for one and eight-tenths. . . .
>
> **Teacher:** Hmm, what did he say?
>
> **Abby:** He said for part of the number line she took part of it, so if I erased all of this, this is what she would have. . . .
>
> **Teacher:** Stop for a moment. What did he say? Is he correct?
>
> **Abby:** I don't know. I'm still stuck. I don't understand. Both of us are kind of at the same point. We both understand, we are saying the same thing, it's just like we don't know if it's a tenth or a hundredth.

The teacher initiates four additional opportunities for students to make connections to the ideas of their peers, but students' responses do not make an explicit connection back to the original student. Instead, students provide a more general mathematical relationship or indicate their own thinking:

> **Teacher:** Why does he think the seven-tenths should be there, Cesar? (1:08)
>
> **Cesar:** Because the seven-tenths should be closer to one than closer to zero. . . .

> **Teacher:** Why does he say that, Lina? (1:16)
>
> **Lina:** Because seven-tenths is greater than five-tenths. . . .
>
> **Teacher** (revoicing): There's no number you can find between three-tenths and four-tenths. Joseph, what do you think about that? (3:15)
>
> **Joseph:** It was confusing with me too but then I thought it could be zero and three and one-half . . .
>
> **Teacher:** Abby, do you agree with that?
>
> **Abby:** I don't really know. I thought that, um, zero and three-tenths would be zero and pretty much three-tenths would be thirty.

With three strong instances of teacher's linking and other consistent attempts to provide opportunities for students to connect to the ideas of their peers, the lesson rates a level 4 on the Teacher's Linking rubric.

TEACHER'S PRESS

The lesson rates a level 4 on the Teacher's Press rubric. The teacher makes the following strong press actions:

> **Teacher:** Okay, what do you mean by that? (1:00)
>
> **Joseph:** 'Cause that's very small and the one is all the way over here and the seven-tenths should be right here (points to a place closer to 1). . . .
>
> **Teacher:** So, Lina, I see that you have made a mark between three-tenths and four-tenths and now you've paused. What are you thinking? (3:00)
>
> **Lina:** I don't know what to put between there. There is no number between it. . . .
>
> **Teacher:** Lina, where are we now? (4:41)
>
> **Lina:** I think that zero and three-tenths is equivalent to zero and thirty-hundredths.

The teacher uses several discourse actions to ask students to clarify units (tenths or hundredths), which in general might serve as press for accurate knowledge but in this problem also elicits students' underlying thinking and understanding. For example, between timestamps 3:37 and 4:07, the teacher continues to ask Abby to clarify whether the "thirty" is tenths or hundredths.

Use the IQA rubrics to identify feedback you would provide to the teacher.

This lesson rates highly on all of the IQA rubrics, and the feedback to the teacher would acknowledge the cognitively demanding nature of the task and the previously identified features of the teacher's implementation and discourse actions that contributed to the high ratings. However, as with all lessons, there are always aspects of instruction that teachers can improve. For example, in the Decimals on a Number Line lesson, a missed opportunity for teacher's press occurs with Abby (3:38). Abby states that she does not know if three-tenths is the same as thirty-hundredths. After this statement, other students make strong links to Abby's contribution. However, the teacher neglects to return to Abby and press for her understanding. We would suggest the teacher return to Abby and ask, "Now what do you think about whether three-tenths is the same as thirty-hundredths?"

Figure 6.2 provides the IQA summary sheet of the ratings for the Decimals on a Number Line lesson.

Dimension	Level	Rationale
Potential of the Task	3	Students can use the number line to make sense of the relationship between tenths and hundredths. There is no explicit prompt for students' reasoning (as required for level 4).
Implementation of the Task	4	Students' reasoning about the placement of decimals on the number line is explicit in the discussion. In the lesson, students explain how they are making sense of the placement of tenths and hundredths.
Teacher's Questions	3	Because most of the time is spent making sense of students' work, we do not hear the teacher asking many original questions. Most of the time, the teacher is pressing students for explanations or to make sense of each other's work.
Teacher's Linking	4	The teacher consistently provides opportunities for students to link to each other's ideas (seven instances). In four instances, students do not make a link but instead explain their own thinking.
Students' Linking	4	Students link to each other's thinking three times when the teacher prompts them and two times on their own.
Teacher's Press	4	There are three strong instances of press and several actions to clarify units (tenths or hundredths), which in this problem also gets at their underlying thinking and understanding.
Students' Providing	4	Students consistently explain their reasoning. There are three strong explanations of the first part of the task (estimate the location of 0.7) and two strong explanations of the third part of the task (decimal between 0.3 and 0.4).

Figure 6.2: IQA summary sheet for the Decimals on a Number Line lesson.

The Decimals on a Number Line lesson portrays the teacher working with a small group of students. This lesson provides examples of strong discourse actions as evidenced by ratings of level 3 or 4 on many of the IQA rubrics.

In activity 6.2, we consider how to use the set of IQA rubrics for a lesson consisting of a whole class engaging in a task, with the teacher interacting with students as they work in small groups and during whole-group discussions.

Activity 6.2: Rating a Whole-Class Lesson

In the following activity, we ask you to rate a whole-class lesson using all the IQA rubrics. It is valuable to engage with tasks as learners prior to considering how they are implemented in a lesson. Be sure to devote attention to this experience. Explore the task on your own before discussing your experience with others.

Engage

In activity 6.2, we visit a middle school mathematics classroom (grade 7) engaged in solving the Bridge Pattern task as seen in figure 6.3 (Nolan, Dixon, Roy, & Andreasen, 2016).

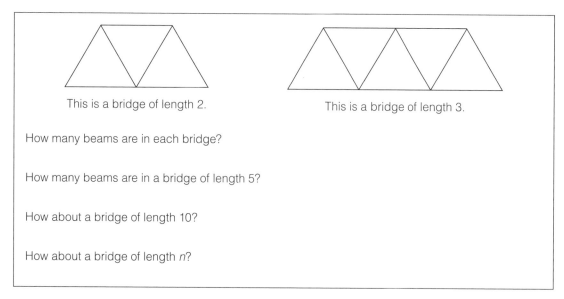

This is a bridge of length 2.

This is a bridge of length 3.

How many beams are in each bridge?

How many beams are in a bridge of length 5?

How about a bridge of length 10?

How about a bridge of length n?

Source: Nolan, Dixon, Roy, & Andreasen, 2016, p. 76.

Figure 6.3: The Bridge Pattern task.

- Solve the Bridge Pattern task using multiple representations and multiple strategies, and discuss your strategies with your collaborative team.

- Find more than one expression to represent the number of beams in a bridge of length n. Use the bridge diagram to justify any expressions you find. How do the numbers and variables in the expression relate to the physical structure of the bridge?

- Before watching the lesson, rate the Bridge Pattern task using the Potential of the Task rubric.

- Watch the Bridge Pattern lesson once the whole way through. Then, replay the clip, stopping the video as needed to take notes and identify examples (with timestamps). As you watch the lesson, take notes and identify examples of discourse actions (and indicate timestamps) that provide instances of Teacher's Questions, Teacher's Linking, Teacher's Press, Students' Linking, and Students' Providing. Because these rubrics are based on a count of discourse actions, it is important to identify specific examples.

Bridge Pattern Lesson:
www.SolutionTree.com/Making_Sense_of_Models

- After watching the video, rate the lesson using the Implementation of the Task rubric (page 33) and use the Implementation Observation Tool (page 47) to support your rating. Rate the Teacher's Questions, Teacher's Linking, Teacher's Press, Students' Linking, and Students' Providing based on the examples of discourse actions you identified.
- Provide a rationale for your rating of each rubric. You can use the IQA summary sheet in figure 6.1 (page 108) as a template. Use the IQA rubrics to identify feedback you would provide to the teacher.

Discuss your ratings, rationales, and feedback with your collaborative team before moving on to the activity 6.2 discussion.

Discuss

How do your responses compare with those in your collaborative team? What themes emerged during your discussion? In this section, we present ideas for you to consider.

Using your notes on the lesson, rate the lesson using all of the IQA rubrics.

In the following sections, we present our ratings for each rubric.

Potential of the Task

The Bridge Pattern task rates a level 3 on the Potential of the Task rubric. The pattern context provides the opportunity for students to use multiple representations (toothpicks, tables, graphs, and numeric or symbolic expressions). There are multiple expressions that could be used to express the general pattern, such as $4n - 1$, $4(n - 1) + 3$, $n + 2n + (n - 1)$, and $3n + (n - 1)$, and each can be connected to a different way of describing the pattern and how it is growing. The task provides an opportunity for students to make sense of variables, constants, and terms in an expression. However, the task does not explicitly prompt students to explain their thinking and reasoning about how they developed the generalization for the bridge of length n.

Implementation of the Task

In the Bridge Pattern lesson, students approach the task using a variety of strategies and representations. Some students use toothpicks, others draw pictures, and others make tables. When students indicate to the teacher that they used the picture or toothpicks to verify the numbers in the table, they show they are making connections between representations. At the end of the lesson, one student describes how the expression relates to the physical structure of the pattern.

However, the students do not work to determine or justify why their observations or generalizations are valid. We do not hear students in small groups explaining why the "adding 4" pattern appears to work, or how $4n - 1$ connects to the diagram. For these reasons, Implementation of the Task rates a level 3. This rating indicates that students *identify patterns but do not justify generalizations*, they *make conjectures but do not provide mathematical evidence or explanations to support conclusions*, and they *use multiple strategies or representations but connections between different strategies or representations are not explicitly evident*.

Several points from section A on the Implementation Observation Tool seem to characterize the lesson. Students *communicate mathematically with peers* and *have access to resources that support their engagement with the task.* The teacher *supports students to engage with the high-level demands of the task while maintaining the challenge of the task, provides sufficient time to grapple with the demanding aspects of the task,* and *provides encouragement for students to make conceptual connections.* While the demands of the task are maintained, we do not have explicit evidence of the items in section C—namely, students *identifying patterns or making conjectures, predictions, or estimates that are well grounded in underlying mathematical concepts or evidence.*

Teacher's Questions

In the Bridge Pattern lesson, the teacher consistently asks original questions, and the lesson rates a level 4 on the Teacher's Questions rubric. Some examples include the following:

- **Teacher:** So, I see you have a table. Can you talk to me about it as well? I see you are talking to your group about it. (1:23)

- **Teacher:** So, how could you figure out a bridge of length ten without having to write the whole table to get to ten? (1:57)

- **Teacher:** How do we get from one term to the next here? How do we get from the three to the seven? (4:03)

- **Teacher:** Does that work? Why don't you talk about this with your group and test it to see if you think it works? (5:11)

- **Teacher:** So where did the four come from? (5:59)

Teacher's Linking

In the lesson, the teacher makes one strong link (2:24): "So you think yes, and you think no. Now you have something interesting to talk about." There are several other linking actions made by the teacher, including revoicing, that support or maintain implicit ways of building on other students' thinking but do not result in students explicitly connecting to the ideas of their peers. For example, at timestamp 1:48, the teacher says, "Now I see you used toothpicks, so you may have been able to check what she did. Did that work?" The student responds, "Yeah," but there is no indication of how he used or understood the idea offered by his peer. A similar discourse action occurs at 4:46 ("Interesting. Check it with a few of yours and make sure you agree"). Additionally, the teacher asks twice (2:18), "Do you agree?" to which students respond *yes* or *no* and do not explain why they agree or disagree. There are instances of revoicing, such as at 4:09 ("Adding four. I wonder if adding four can help us here?") as well as 6:04 and 6:21. With one strong link, plus revoicing and several other links that contributed to implicit connections, the lesson rates a level 2 on the Teacher's Linking rubric.

Students' Linking

In the lesson, there is one instance of strong students' linking. This action occurs after the teacher makes a strong linking action and walks away:

Teacher: So you think yes, and you think no. Now you have something interesting to talk about. *(Teacher walks away.)* (2:24)

> **Student** (to his groupmate): How'd you get the conclusion that it was twenty? (2:31)

There are other instances of implicit connections throughout the lesson, where students are using each other's tables or results or stating that they agree or disagree with a classmate, but without providing an explicit connection to the other student's thinking or an explanation of why they agree or disagree. For example, during the exchange beginning at 4:03, the teacher asks, "How do we get from one term to the next here? How do we get from the three to the seven?" A student responds, "Adding four," and other students use the idea of adding four to respond to the next few questions asked by the teacher (for example, "What about seven to eleven? What are we doing?"). However, students' understanding of the idea of "adding four" is not clear, as they do not explain why the pattern grows by four each time or where the groups of four are represented in the pattern. They also do not make a clear reference back to the original student. For this segment to contain a strong link, students would need to say, "He said adding four because . . ." or something similar. The exchange beginning at 4:29 also demonstrates an implicit connection, without explicit links between students. Hence, the lesson rates a level 2 on the Students' Linking rubric.

Teacher's Press

The teacher presses students for conceptual explanations in the following instances:

> **Teacher:** So, I see you have a table. Can you talk to me about it as well? I see you are talking to your group about it. (1:23)
>
> **Teacher:** So, you think a bridge of length ten would be thirty-five? (2:15)
>
> **Teacher:** Another segment? Another beam here? (points to the diagram) (6:04)

The students' explanations that follow these discourse actions do not always provide evidence of students' thinking and reasoning. Following the teacher's press at 2:15 and 6:01, the students express their agreement, and the discussion moves on. There are also examples of missed opportunities for press, in which a student offers an important idea and the conversation moves on without an explanation or unpacking of the idea:

> **Student:** We tried it with the numbers in the problems we did earlier, and we got the same results. (5:27)
>
> **Teacher:** Interesting. So, who did it another way?
>
> **Student:** I did $4n - 1$. (5:36)
>
> **Teacher:** How many of you did it like that?

For these reasons, the lesson rates a level 3 on the Teacher's Press rubric. There are instances of pressing for conceptual understanding, but such discourse actions are not consistent throughout the lesson.

STUDENTS' PROVIDING

The lesson rates a level 4 on the Students' Providing rubric. Throughout the lesson, students often provide evidence for their claims or offer explanations that indicate their understanding:

> **Student:** I made the table because, since the length of the first bridge is two, so I wrote it down as x, and the beam is seven. . . . Then the second bridge, the length is three and the beam is eleven. Then, I predicted that, each time it's just adding four to the beams. (1:30)
>
> **Student:** I thought, if you added another five times the four beams, it would give you twenty, and I added twenty to nineteen, and it gives me thirty-nine. (2:41)
>
> **Student:** One triangle has three sides, and then they added another one. (6:01)
>
> **Student:** I just thought since, because it is minus one since the last one didn't have another line on it. (6:15)

While there are explanations and justifications of other mathematical ideas and relationships that we would like to have heard (as identified previously in the discussion of the Implementation of the Task rubric or Teacher's Press rubric), in the students' responses we do hear, students consistently provide conceptual explanations.

Figure 6.4 provides a summary sheet for the Bridge Pattern lesson. How might you use the data to provide feedback that informs instructional practice? The following section discusses how the IQA rubrics can serve as a tool for reflecting on, providing feedback on, and improving instruction.

Dimension	Level	Rationale
Potential of the Task	3	There is opportunity for using multiple representations, using multiple expressions, and making sense of variables, constants, and terms in an expression.
Implementation of the Task	3	Students use a variety of strategies and representations and appear to begin to make connections. Students do not provide justifications for their conjectures and expressions or explicit evidence of connections between representations.
Teacher's Questions	4	The teacher consistently asks academically relevant questions.
Teacher's Linking	2	The teacher provides one strong link plus prompts implicit connections and revoicing.
Students' Linking	2	Students provide one strong link plus implicit use of each other's ideas.
Teacher's Press	3	There are some instances of press for conceptual understanding, but not consistent press, as there are also instances of missed press.
Students' Providing	4	Students provide explanations of their thinking, sometimes without the teacher prompting them.

Figure 6.4: IQA summary sheet for the Bridge Pattern lesson.

The IQA Rubrics as Tools to Reflect on and Improve Instruction

Imagine that the teacher of the Bridge Pattern lesson shared her video with her collaborative teammate, and they both reflected on the lesson using the IQA rubrics. We provide an example of a possible teacher's reflection in figure 6.5.

> As I watched the video, I was pleased with the discussion within the first group, even after I left the group (2:31). I also thought that asking students to work on bridges of lengths five and ten and then calling the whole group back together briefly (2:55–3:55) to give the additional direction for a bridge of length n was helpful in supporting students' learning and access to the task.
>
> When I called the group back together, while it seemed strange not to press the student's answer of thirty-nine (3:18), I think I would make the same decision again in a future situation. This was still the beginning of the lesson, and I intentionally did not ask the student how she found thirty-nine beams because I didn't want to give away a strategy to the rest of the class. I still think that was a good choice because I wanted the students to find a general strategy and pattern. I saw many students simply extending the "adding four" (recursive) pattern to find the number of beams for bridges of lengths five and ten, and I wanted them to do more than that. For a bridge of length n, they would need to find a functional relationship between the bridge length and the number of beams—I was hoping that they would begin to reason that way with the bridge of length n.
>
> My collaborative teammate, who reviewed the video with me, asked if I would choose to say, "We could set n equal to the length of the bridge. What expression could we come up with?" (3:49) if I taught the lesson again. She wondered whether that statement limited students' opportunities for problem solving. She pointed out the following lines from the Implementation Observation Tool (figure 2.9, page 47):
>
> > *provided a set procedure for solving the task, and gave feedback, modeling, or examples that were too directive or did not leave any complex thinking for the student*
>
> I agree that telling students to set n equal to the length of the bridge and find an expression provides students with a general procedure for solving the task, so perhaps this statement did take away some opportunities for problem solving. However, based on the variety of strategies I observed as students worked, the task did not seem to become proceduralized.
>
> A few things stood out to me when I watched the video that I would consider changing. Twice, I asked students to "test" the numbers in an expression to determine if the expression "worked" (timestamps 4:45 and 5:13). While this is helpful to notice if the conjecture is reasonable, mathematically, proving whether an expression worked (or was valid) would require testing every possible number. I worry that I might have led students to a misconception that you can prove by example. Next time, I would suggest that students explore their conjectures with a few examples, and I would also encourage them to justify the expressions using the diagram—the pattern and how it is growing, and the connection between the bridge length and the number of beams. As I watched the video, I wanted to hear more of that in students' explanations. For example, if students have the expression $n + 2n + (n - 1)$, they might explain that a bridge of length n has a base of n beams, a middle of $2n$ beams, and a top of $n - 1$ beams. Asking for this type of justification would have made students' thinking more explicit and could raise the rating on the Implementation of the Task rubric to a level 4. Two students provide this type of explanation at the end of the video when I ask them to make sense of $4n - 1$ using the picture.
>
> Another thing we noticed is that I asked a lot of questions. My colleague and I both rated Teacher's Questions as a level 4. However, I didn't use much press and even less linking—maybe because I kept asking new questions rather than questions that connected to what students said. For example, the exchange with students beginning at 4:14 ("What about seven to eleven? What are we doing?") and continuing through 4:22 ("So how can we use that in our expression?") sounded more like a series of short questions rather than pressing students' ideas or knowledge, even though I was pointing to the numbers in the student's chart. I felt the same about 4:29 ("So what happens if a bridge is of length two? What's two times four?") through 4:40 ("But what do we need?"). Watching the video, I felt like I was asking questions for what I wanted students to do or notice rather than having them unpack their own thinking. I rated Teacher's Press a level 3 for that reason.

> This was a hard lesson for me to teach. I think I did pretty well, but after reflecting using the Implementation Observation Tool, I see how I can make relatively small changes that will push students to a deeper level of reasoning. The next time I implement a pattern task, I am going to plan discourse actions that ask students to relate their expressions back to the diagram. Then I will be able to hear how students are making sense of the variables, constants, coefficients, and operations used in their expressions.

Figure 6.5: Sample teacher's reflection on the Bridge Pattern lesson.

Notice how the teacher used all the rubrics in the IQA Toolkit to assess various elements of her instructional practice and concluded by identifying specific things about her practice that she can work on improving. This teacher used the entire IQA Toolkit to reflect on her instructional quality; however, each rubric in the IQA Toolkit can be used separately to focus on specific areas of instructional improvement.

In the following section, the application activities will guide you to your next steps as you begin to use the IQA Toolkit to assess your own instructional practices and those of your collaborative team.

Application Activities

In these application activities, you will practice using the IQA Toolkit to inform instructional quality.

Activity 6.3: Considering the Teacher's Reflection and Next Steps

In activity 6.3, we provide the opportunity to consider the teacher's reflection and provide feedback to the teacher.

Engage

Discuss the teacher's reflection in figure 6.5 with your collaborative team. Identify at least one place in the lesson where a different question, link, or press could have served to support students to provide justifications, explanations, or connections between representations. Write the timestamp and the exact wording of the question, link, or press you would use.

Discuss

How do your responses compare with those in your collaborative team? What themes emerged during your discussion? In this section, we present ideas for you to consider.

Identify at least one place in the lesson where a different question, link, or press might have supported students to provide justifications, explanations, or connections between representations.

There are some instances in which the teacher might have referred students back to their drawing or models of the bridge rather than to the table of values or expression. For example, beginning at 4:03, the teacher asks a series of short questions about the table of values. Students realize the "adding four" relationship from one value in the table to the next. At any point before the teacher leaves the group at 4:46, she could have asked, "How does that relate to what you see happening in the picture?" Similarly, consider the following exchange:

> **Student:** 'Cause it's always gonna be one extra; all you gotta do is subtract one. (4:41)
>
> **Teacher:** Interesting. Check it with a few of yours and make sure you agree.

The student who speaks at 4:41 has a diagram of the bridge pattern and "$4n - 1 =$" written on her whiteboard, and the other three students in the group all have tables of values on their whiteboards. The teacher's reply is directed to the group of students, indicating that they should check and determine if the numbers in the tables support the idea of multiplying by four and subtracting one. Because a diagram of the pattern and an expression are part of one student's work, the teacher could have replied, "Interesting. See if you can make sense of how $4n - 1$ relates to this diagram."

Consider also this exchange that occurred during the whole-group discussion:

> **Teacher** (calls class back together): What did you find out? (5:24)
>
> **Student:** We tried it with the numbers in the problems we did earlier, and we got the same results.
>
> **Teacher:** Interesting. So, who did it another way?

The teacher might have chosen to press the student to say more about what it means to have "tried it with the numbers in the problems." The student is referring to the expression $4(n - 1) + 3$. The teacher could also have asked, "Is there a way to determine if the expression will always work, for any number and not just for the numbers in your tables?" or she could have explicitly asked students to make sense of the expression using the diagram.

You may have noted other instances in which a question, link, or press could have elicited students' making connections between representations, making sense of the expression, or providing a justification using the diagram. As the teacher noted in her reflection, however, sometimes there is a valid reason not to press an idea further, namely, if sharing too much explanation with the whole class (or group) would take away other students' opportunities for thinking, reasoning, and problem solving. At other times, the student's idea or strategy might cause the class to veer away from the main mathematical ideas of the lesson, or the teacher may be trying to use the limited time in the lesson to provide space for other students to contribute their ideas to the discussion.

In general, when reviewing a lesson and providing feedback, or when planning discourse actions to use in a lesson, consider including discourse actions that:

- Elicit students' ideas about the main mathematical goals of the lesson
- Provide opportunities for students' voices to be heard
- Provide opportunities for the teacher to hear evidence of students' thinking and reasoning

While some additional discourse actions may have contributed to several rubrics rating a level 4, including Implementation of the Task, the ways students engage with the task (and the way the teacher supports students to engage with the task) served to maintain the rating on the Implementation of the Task rubric at a level 3. The task rated a level 3 on the Potential of the Task rubric, and so the teacher enacted the task during the lesson at the same level at which it appeared in print—in other words, the task directions did not include a prompt to connect the expression to the diagram or provide a justification. This lesson is typical of lessons that rate a level 3 on the Implementation of the Task rubric. In such lessons, students engage in problem solving and reasoning in solving a task, and students' opportunities to engage with the task are not reduced to merely following rote procedures (level 2) or memorization (level 1), yet students' thinking and reasoning are not explicitly evident in their work or discussion. While you should aim to implement tasks at a level 4, sometimes the constraints of teaching, including pressures regarding time, might warrant remaining at a level 3.

This lesson also illustrates the typical relationship among Teacher's Questions, Teacher's Linking, and Teacher's Press discussed earlier in this chapter. The teacher asks several good questions and uses some instances of press, but makes fewer links. Small changes in questions, linking, and press can make students' thinking explicit, public, accessible by their peers, and assessable by their teacher, and thus increase the rating to a level 4 on the Teacher's Questions, Teacher's Linking, and Teacher's Press rubrics.

Activity 6.4: Revisiting the Chapter 5 Transition Activity—Using All IQA Rubrics

In the chapter 5 transition activity (page 105), you observed and analyzed mathematics lessons using all of the IQA rubrics. Hopefully, you have had a chance to watch and rate your own lesson and a lesson from someone within your collaborative team.

Engage

Pair up with a member of your collaborative team to discuss and compare your examples and ratings for each lesson.

- Compare and discuss your ratings for Potential of the Task and Implementation of the Task. Identify criteria from the rubrics and from the Implementation Observation Tool that contributed to your rating. Discuss how the task and implementation provide access for all students.
- Share your examples of teacher's questions. Compare and discuss your rating of the Teacher's Questions rubric and your justifications for the rating.
- Share your examples of teacher's linking and teacher's press. Compare and discuss your rating and the reasoning for the rating on the Teacher's Linking and Teacher's Press rubrics.
- Share your examples of students' linking and students' providing. Compare and discuss your rating of the Students' Linking and Students' Providing rubrics. Be sure to include your justifications.
- What did you notice about students' access in the lesson? How did the rubrics support the examination of each and every student having access to the task?

Based on your notes and ratings, decide on some next steps for instructional improvement.

Discuss

How do your responses compare with those in your collaborative team? What themes emerged during your discussion? In this section, we present ideas for you to consider.

Pair up with a member of your collaborative team to discuss and compare your examples and ratings for each lesson.

How consistent were your examples and ratings? Did you find it helpful to hear what a colleague considers as strong discourse actions or students' responses? Teachers with whom we have worked have found it helpful to use the rubrics to review another teacher's lesson, using the specific indicators in the rubric levels to point to when constructively critiquing another teacher's lesson (Boston & Candela, 2018). The rubrics also provide a common language for discussions about mathematics instruction among teachers and between teachers and administrators (Boston, Henrick, Gibbons, Berebitsky, & Colby, 2016) and a consistent set of practices for teachers to work toward over time.

Based on your notes and ratings, decide on some next steps for instructional improvement.

What did the rubrics and ratings indicate about next steps for instructional improvement? In any rubric, level 3 may not signal the need for improvements, or it may indicate where the planning of small changes or more frequent use of certain practices might be warranted. Deeper consideration of instructional practices is necessary when a task is rated at level 3 or 4 on the Potential of the Task rubric and falls to level 1 or 2 on the Implementation of the Task rubric. In these cases, students' work during implementation focuses on procedures or memorization rather than sense making and reasoning, contrary to the design of the task. As an action plan in these instances, identify specific components of the rubrics or Implementation Observation Tool that seem to characterize the lesson, and use those to plan the enactment of a future level 3 or 4 task.

For Teacher's Questions, Teacher's Linking, or Teacher's Press, level 2 would signal the need for discourse actions that elicit students' understanding and enable students to make connections to the ideas of their peers. Sometimes, a level 2 in one of these three rubrics might be warranted if the other two rubrics are rated a level 3 or 4, as the nature of the discussion or mathematical topic may promote some questions, linking, or press over others. For example, a lesson in which a teacher has students debating two ideas may elicit more linking, particularly when one idea is a common misconception. However, level 1 in any of these rubrics would highlight an area in need of improvement. Fortunately, teachers can plan strong discourse actions ahead of time so they are ready to go when needed in a lesson. As next steps, when planning a lesson, use the ideas, examples, and rubrics in this book (as well as other examples in the *Making Sense of Mathematics for Teaching* series) to craft discourse actions that support students' thinking, reasoning, and connections.

The level of the Students' Linking and Students' Providing rubrics offers a different window into a lesson—rather than indicating teachers' practice, they provide evidence of students' engagement. As described in chapter 5, improving instructional practice around the other rubrics should enhance students' opportunities to link to the ideas of their peers and provide explanations and justifications. Using linking and press, you can support students' engagement by helping them link to a classmate's thinking or ask a question to clarify a student's thinking and support his or her understanding of the task. We have found that students in classrooms with levels 3 and 4 ratings for Students' Linking and Students' Providing tend to rate their learning experiences in those classrooms as more enjoyable, equitable, and empowering than in classrooms where the Students' Linking and Students' Providing rubrics rate level 1 or 2 (Parke, Boston, Thomas-Browne, & Orsini, in press). In the introduction, we noted that often feedback around instruction will include general comments about students' engagement. Now, you can use the rubrics in chapter 5 to guide and focus your analysis, reflection, and feedback about specific aspects of students' engagement.

What did you notice about students' access in the lesson? How did the rubrics support the examination of each and every student having access to the task?

If you used a task at a level 3 or 4 on the Potential of the Task rubric, you may have noticed that because the task has multiple entry points and multiple solution methods, it provided each and every student an opportunity to engage in the task, both within peer groups and in the whole-class discussion. Tasks at a level 1 or 2 for Potential of the Task limit access by requiring students to know specific procedures, facts, rules, or formulas from memory. It is an interesting paradox that tasks with lower levels of cognitive demand—tasks that rate a level 1 or 2 on the Potential of the Task rubric—are often more difficult and less supportive of students who struggle with mathematics. Tasks with higher levels of cognitive demand—tasks that rate a level 3 or 4 on the Potential of the Task rubric—often have fewer barriers and greater access in allowing multiple ways to approach the task, use of multiple representations or modeling, and connections to real-life contexts or prior mathematical knowledge.

Lessons with higher ratings for Teacher's Questions, Teacher's Linking, and Teacher's Press also promote greater access by including more students' voices in the lesson. As you plan discourse actions and the type of evidence you want to collect, consider how to bring each and every student into the lesson. As teachers, we make decisions about which students have opportunities to talk, which students to ask particular questions, and which students to give space to explain their ideas or link to the ideas of their peers. We can use discourse actions in the classroom to raise students' confidence in themselves and their status among their peers as learners and doers of mathematics. By attending to who has opportunities to answer questions, link to peers, or provide justification, teachers can promote equitable access to high-quality learning opportunities. Finally, being purposeful about who is asked to answer questions, link, or provide an explanation during a lesson can increase students' sense of agency or belief that they can think, create, and enact mathematics rather than merely reproducing mathematics that someone else has shown them.

Summary

Now that you've had a chance to use the set of rubrics to examine your own practice or that of your collaborative team, take a moment to reflect on how you see yourself integrating these rubrics on a regular basis. We invite you to use the rubrics to examine your selection and implementation of cognitively demanding tasks, your discourse actions, and student responses to support student engagement and increase student achievement.

EPILOGUE
Next Steps

Throughout this book, you have focused on instructional practice by considering the nature of mathematical tasks, implementation, teacher's questions, teacher's linking and press, and students' responses. Often, as teachers and schools engage in professional learning experiences, instructional improvements and change in teachers' practices follow this progression from tasks to overall implementation to discourse actions (Boston & Smith, 2009). As teachers make these changes, the nature of students' linking and providing changes as well.

Now that you have used the set of IQA rubrics to assess practice, you may want to move beyond assessing individual lessons and instead analyze sets of lessons, or periodically assess lessons over time. If you are a classroom teacher, we encourage you to continue to use the IQA rubrics to self-assess and reflect on your instructional practice. Ideally, find a peer and engage in reciprocally observing or video recording lessons in each other's classrooms. Rate and discuss the lessons and use the rubrics to identify next steps, monitor progress, and celebrate accomplishments in instructional practice. We suggest that you use the rubrics to reflect on instruction several times throughout the year. You might decide to use the entire set of IQA rubrics holistically, or you may choose to use individual rubrics that align with your goals for instructional improvement.

If you serve as an instructional leader, you may also want to use the IQA rubrics to collect data on groups of teachers, either as a checkpoint or to assess progress over time. The IQA rubrics can provide data to inform efforts in professional development or curriculum implementation at the school or district level. The IQA Toolkit, and this book, can also serve as the basis of professional development efforts, as a framework for considering instructional practices necessary for implementing a new curriculum effectively, or as next steps in instructional change following teachers' exploration and deeper learning of specific mathematics topics.

As we describe in the introduction, the set of IQA rubrics provides a framework of instructional practices that can serve to focus instructional growth and change. You might choose to continue to reflect on instruction using the entire set of rubrics, as we did in chapter 6, or to select specific rubrics to focus your ongoing reflections. The added value of using the IQA rubrics comes in the levels within each rubric, and the specific descriptions of each of those levels. In each rubric, changes in the instructional practices described from one level to the next can suggest pathways for instructional improvement. If you plan to use the IQA rubrics to collect data reliably over time, we encourage you to spend additional time reaching consensus on the rating of tasks and lessons within your collaborative team. Appendix D (page 141) provides additional guidance for rating instructional tasks.

While it may initially be uncomfortable to assign a numerical rating to the teaching practice of yourself or a colleague, we noted in the introduction that the ratings and levels have several important purposes.

First, with the detailed descriptions of each level, the numerical ratings provide a strong descriptive sense of the instructional practice in each rubric. After exploring this book, you can probably describe what questioning at a level 2 on the Teacher's Questions rubric looks like in a mathematics classroom, or what an example of a student response sounds like if it qualifies as a strong link toward a rating of level 4 on the Students' Linking rubric. Second, the descriptions and levels provide pathways for improvement by indicating what is necessary to advance to the next level. Third, the descriptions and levels provide a common language for reflections, conversations, and feedback. Rather than having to craft feedback for a member of your collaborative team, you can use language and ideas directly from the rubrics. Finally, numerical ratings provide a condensed way to collect and record data over time or among groups of teachers. While this feature may be more useful in research (see, for example, Boston, 2012; Boston & Candela, 2018; Boston & Smith, 2009; or Boston & Wilhelm, 2015, for more detail on using the IQA Toolkit in classroom observation research), it may also provide data to support continued growth among individuals or teams of teachers.

As this book comes to a close, we encourage your work around watching, sharing, rating, and reflecting on lessons with your collaborative team to be ongoing. Continue to observe or video record mathematics lessons, and use the IQA rubrics to rate the lessons, identify feedback, and determine pathways for instructional improvements. In other words, continue to engage in the chapter 5 transition activity (page 105) and activity 6.4 (page 121). We hope you and your collaborative team find the IQA Toolkit useful in providing a framework to assess mathematics instruction, language to express and focus feedback, and pathways to drive instructional improvements.

APPENDIX A
The IQA Toolkit

IQA Potential of the Task Rubric

4	The task has the potential to engage students in complex thinking or in creating meaning for mathematical concepts, procedures, or relationships.
	The task *must explicitly prompt* for evidence of students' reasoning and understanding. For example, the task may require students to:
	• Solve a genuine, challenging problem for which students' reasoning is evident in their work on the task
	• Develop an explanation for why formulas or procedures work
	• Identify patterns and form and justify generalizations based on these patterns
	• Make conjectures and support conclusions with mathematical evidence
	• Make explicit connections between representations, strategies, or mathematical concepts and procedures
	• Follow a prescribed procedure in order to explain or illustrate a mathematical concept, process, or relationship
3	The task has the potential to engage students in complex thinking or in creating meaning for mathematical concepts, procedures, or relationships.
	However, the task does not warrant a level 4 rating because it does not explicitly prompt for evidence of students' reasoning and understanding. For example, students may be asked to:
	• Engage in problem solving, but for a task that provides minimal cognitive challenge (for example, a problem that is easy to solve)
	• Explore why formulas or procedures work, but not to provide an explanation
	• Identify patterns, but not to explain generalizations or provide justification
	• Make conjectures, but not to provide mathematical evidence or explanations to support conclusions
	• Use multiple strategies or representations, but not to develop connections between them
	• Follow a prescribed procedure to make sense of a mathematical concept, process, or relationship, but not to explain or illustrate the underlying mathematical ideas or relationships
2	The potential of the task is limited to engaging students in using a procedure that is either specifically called for, or its use is evident based on prior instruction, experience, or placement of the task.
	• There is little ambiguity about what needs to be done and how to do it.
	• The task does not require students to make connections to the concepts or meaning underlying the procedure they are using.
	• The focus of the task appears to be on producing correct answers rather than developing mathematical understanding (for example, applying a specific problem-solving strategy or practicing a computational algorithm).
1	The potential of the task is limited to engaging students in memorizing; note taking; or reproducing facts, rules, formulas, or definitions. The task does not require students to make connections to the concepts or meanings that underlie the facts, rules, formulas, or definitions they are memorizing or using.

Source: Adapted from Boston, 2017.

IQA Implementation of the Task Rubric

4	Students engaged in exploring and understanding the nature of mathematical concepts, procedures, or relationships. There is *explicit evidence* of students' reasoning and understanding. For example, students may have: • Solved a genuine, challenging problem for which students' reasoning is evident in their work on the task • Developed an explanation for why formulas or procedures work • Identified patterns and formed and justified generalizations based on these patterns • Made conjectures and supported conclusions with mathematical evidence • Made explicit connections between representations, strategies, or mathematical concepts and procedures • Followed a prescribed procedure in order to explain or illustrate a mathematical concept, process, or relationship
3	Students engaged in complex thinking or in creating meaning for mathematical concepts, procedures, or relationships. However, the implementation does not warrant a level 4 because there were *no explicit explanations or written work* to indicate students' reasoning and understanding. Students may have: • Engaged in problem solving, but for a task that required minimal cognitive challenge (for example, the problem was easy to solve), or students' reasoning is not evident in their work on the task • Explored why formulas or procedures work but did not provide explanations • Identified patterns but did not form or justify generalizations • Made conjectures but did not provide mathematical evidence or explanations to support conclusions • Used multiple strategies or representations but connections between different strategies or representations were not explicitly evident • Followed a prescribed procedure to make sense of a mathematical concept, process, or relationship, but did not explain or illustrate the underlying mathematical ideas or relationships
2	Students engaged in using a procedure that either was specifically called for or its use was evident based on prior instruction, experience, or placement of the task. • There was little ambiguity about what needed to be done and how to do it. • Students did not make connections to the concepts or meaning underlying the procedure being used. • The focus of the implementation appeared to be on producing correct answers rather than developing mathematical understanding (for example, applying a specific problem-solving strategy or practicing a computational algorithm).
1	Students engaged in memorizing; note taking; or reproducing facts, rules, formulas, or definitions. Students did not make connections to the concepts or meanings that underlie the facts, rules, formulas, or definitions being memorized or reproduced.
0	Students did not engage in mathematical activity.

Source: Adapted from Boston, 2017.

IQA Teacher's Questions Rubric

4	The teacher consistently asks academically relevant questions that provide opportunities for students to elaborate and explain their mathematical work and thinking (*probing students' thinking* and *generating discussion*); identify and describe the important mathematical ideas in the lesson; or make connections among ideas, representations, or strategies (*exploring mathematical meanings and relationships*).
3	At least three times during the lesson, the teacher asks academically relevant questions that provide opportunities for students to elaborate and explain their mathematical work and thinking (*probing students' thinking* and *generating discussion*); identify and describe the important mathematical ideas in the lesson; or make connections among ideas, representations, or strategies (*exploring mathematical meanings and relationships*).
2	The teacher makes limited, superficial, trivial, or formulaic efforts to ask academically relevant questions that provide opportunities for students to elaborate and explain their mathematical work and thinking (*probing students' thinking* and *generating discussion*); identify and describe the important mathematical ideas in the lesson; or make connections among ideas, representations, or strategies (*exploring mathematical meanings and relationships*). For example, the teacher asks every student the same question or set of questions; there are one or two instances of strong questions; or the teacher asks the same strong question multiple times.
1	The teacher asks procedural or factual questions that elicit mathematical facts or procedures or require brief, single-word responses (*eliciting procedures or facts*).
0	The teacher does not ask questions during the lesson, or the teacher's questions are not relevant to the mathematics in the lesson (*inquiring about other mathematical topics* or *asking nonmathematical questions*).

Source: Adapted from Boston, 2017.

IQA Teacher's Linking Rubric

4	The teacher consistently (at least three times) explicitly connects (or provides opportunities for students to connect) speakers' contributions to each other *and* describes (or provides opportunities for students to describe) how ideas or positions shared during the discussion relate to each other.
3	At least twice during the lesson, the teacher explicitly connects (or provides opportunities for students to connect) speakers' contributions to each other *and* describes (or provides opportunities for students to describe) how ideas or positions relate to each other.
2	At one or more points during the discussion, the teacher links speakers' contributions to each other, but *does not show* how ideas or positions relate to each other (for example, implicitly building on ideas; or noting that ideas or strategies are different but not describing how). The teacher may revoice or recap, but *does not describe* how ideas or positions relate to each other, or makes only one strong effort to connect speakers' contributions to each other (one strong link).
1	The teacher does not make any effort to link or revoice speakers' contributions.
0	There is no class discussion, or class discussion is not related to mathematics.

Source: Adapted from Boston, 2017.

IQA Teacher's Press Rubric

4	The teacher consistently (almost always) asks students to provide evidence for their contributions by pressing for conceptual explanations or to explain their reasoning. There are few, if any, instances of missed press, in which the teacher needed to press and did not.
3	At least twice during the lesson, the teacher asks students to provide evidence for their contributions by pressing for conceptual explanations or to explain their reasoning. The teacher sometimes presses for explanations, but there are instances of missed press.
2	Most of the press is for computational or procedural explanations or memorized knowledge, or there are one or more superficial, trivial, or formulaic efforts to ask students to provide evidence for their contributions or to explain their reasoning (for example, asking, "How did you get that?") before then moving on without attending to student responses.
1	There are no efforts to ask students to provide evidence for their contributions, and there are no efforts to ask students to explain their thinking.
0	There is no class discussion, or class discussion is not related to mathematics.

Source: Adapted from Boston, 2017.

IQA Students' Linking Rubric

4	The students consistently, explicitly connect their contributions to each other and describe how ideas or positions shared during the discussion relate to each other (for example, "I agree with Sam because . . .").
3	At least twice during the lesson, students explicitly connect their contributions to each other and describe how ideas or positions shared during the discussion relate to each other (for example, "I agree with Mohammed because . . .").
2	At one or more points during the discussion, the students link students' contributions to each other, but do not describe how ideas or positions relate to each other (for example, implicitly using or building on others' ideas; or "I disagree with Ana"), or students make only one strong effort to connect their contributions with each other.
1	Students do not make any effort to link or revoice students' contributions.
0	Class discussion is not related to mathematics, or there is no class discussion.

Source: Adapted from Boston, 2017.

IQA Students' Providing Rubric

4	Students consistently provide evidence for their claims, or explain their thinking using reasoning in ways appropriate to the discipline (for example, conceptual explanations).
3	Once or twice during the lesson, students provide evidence for their claims, or explain their thinking using reasoning in ways appropriate to the discipline (for example, conceptual explanations).
2	Students provide explanations that are computational, procedural, or memorized knowledge, or what little evidence or reasoning students provide is inaccurate, incomplete, or vague.
1	Students do not back up their claims or do not explain the reasoning behind their claims.
0	Class discussion is not related to mathematics, or there is no class discussion.

Source: Adapted from Boston, 2017.

IQA Implementation Observation Tool

A ↑	The lesson **provides** opportunities for students to engage with high-level cognitive demand.	B ↓	The lesson **does not provide** opportunities for students to engage with high-level cognitive demand.
<ul><li>Students:<ul><li>Engage with the task in ways that address the teacher's goals for high-level thinking and reasoning</li><li>Communicate mathematically with peers</li><li>Have appropriate prior knowledge to engage with the task</li><li>Have opportunities to serve as mathematical authorities in the classroom</li><li>Have access to resources that support their engagement with the task</li></ul></li><li>The teacher:<ul><li>Supports students to engage with the high-level demands of the task while maintaining the challenge of the task</li><li>Provides sufficient time to grapple with the demanding aspects of the task and to expand thinking and reasoning</li><li>Holds students accountable for high-level products and processes</li><li>Provides consistent requests for explanation and meaning</li><li>Provides students with sufficient modeling of high-level performance on the task</li><li>Provides encouragement for students to make conceptual connections</li></ul></li></ul>			<ul><li>The task:<ul><li>Expectations are not clear enough to promote students' engagement with the high-level demands of the task</li><li>Is not rigorous enough to support or sustain student engagement in high-level thinking</li><li>Is too complex to sustain student engagement in high-level thinking (Students do not have the prior knowledge necessary to engage with the task at a high level.)</li></ul></li><li>The teacher:<ul><li>Allows classroom management problems to interfere with students' opportunities to engage in high-level thinking</li><li>Provides a set procedure for solving the task</li><li>Shifts the focus to procedural aspects of the task or on correctness of the answer rather than on meaning and understanding</li><li>Gives feedback, modeling, or examples that are too directive or do not leave any complex thinking for the student</li><li>Does not press students or hold them accountable for high-level products and processes or for explanations and meaning</li><li>Does not give students enough time to deeply engage with the task or to complete the task to the extent that is expected</li><li>Does not provide students access to resources necessary to engage with the task at a high level</li></ul></li></ul>

C	The **discussion** provides opportunities for students to engage with the high-level demands of the task.

Students:

- Use multiple strategies and make explicit connections or comparisons between these strategies, or explain why they chose one strategy over another
- Use or discuss multiple representations and make connections between different representations or between the representation and their strategy, underlying mathematical ideas, or the context of the problem
- Identify patterns or make conjectures, predictions, or estimates that are well grounded in underlying mathematical concepts or evidence
- Generate evidence to test their conjectures and use this evidence to generalize mathematical relationships, properties, formulas, or procedures
- Determine the validity of answers, strategies, or ideas rather than waiting for the teacher to do so

Source: Adapted from Boston, 2017.

Framework for Different Types of Questions

Question Type	Description	Examples From the Father and Son Race Lesson Version 1
Probing students' thinking	• Clarifies student thinking • Enables students to elaborate on their own thinking for their own benefit and for the class	• So what are you doing? • What are we talking about with accelerating? • Time is standing still? Try to explain. • So what is your thinking about this now?
Exploring mathematical meanings and relationships	• Points to underlying mathematical relationships and meanings • Makes links between mathematical ideas	• What was happening at the start of the race? • Where are they when the race starts? • What in the graph tells you that? • What is happening with the speed? • What's happening at B? • Where are they at C? • What does the "up" mean? • Is there a point during this race where the son was running at a different rate?
Generating discussion	• Enables other members of the class to contribute and comment on ideas under discussion	• Do you agree? • So you agree it is faster? • What does she mean by that? • Why does she say "no"? • Agree or disagree? Why?
Eliciting procedures or facts	• Elicits a mathematical fact, procedure, or definition • Requests the result (only) of a mathematical procedure • Requests units or terminology • Requires a yes-or-no or single-word answer	• At four what? • Zero what? • What do we call what is happening at A?
Inquiring about other mathematical topics	• Relates to teaching and learning mathematics, but does not directly relate to the task or mathematical ideas of focus for the lesson	General examples (not from the lesson): • What else could you graph and find the point of intersection? • How else could you label this graph?
Asking nonmathematical questions	• Does not relate to teaching and learning mathematics • May relate to the context of the lesson or task	General examples (not from the lesson): • Do you want to use graph paper? • Who has competed in a race in track?

Source: Adapted from Boston, 2017.

APPENDIX B
Suggested Answers for Activity 1.4

Task	Rationale	Ways to Adapt the Task to Increase the Cognitive Demand
Level 3: Number Pairs That Make 10 Grade K Two-sided chips (red and yellow) are available. Jasmine has 10 marbles. Some of them are red and the rest are yellow. How many marbles could be red and how many marbles could be yellow?	The task provides a context for kindergarten students to model and make sense of number pairs that sum to 10. There are multiple solutions and students can develop and use their own strategy (no strategy is suggested or provided by the task). The task does not explicitly prompt for an explanation. This task provides access as students could use multiple methods to solve and could use manipulatives to model possible solutions.	Add a prompt to explain: • Does more than one number pair make 10? Why? • Provide two different ways to solve the problem, and reflect on or compare the strategies in some way. • Find all of the possible number pairs. How do you know you have found all the possible number pairs? • Draw a picture of two number pairs you have found. How are your pictures the same and how are they different?
Level 2: Adding Fractions With Unlike Denominators Grade 5 In problems 1–3, find a common denominator and add the fractions: 1) $\frac{3}{4} + \frac{1}{20} =$ 2) $\frac{2}{3} + \frac{3}{5} =$ 3) $\frac{5}{12} + \frac{1}{6} =$	The task provides the opportunity for students to practice or demonstrate a previously learned procedure for adding fractions with unlike denominators. Nothing in the task supports students to make sense of the procedure.	Include a context or representation: ask students to determine a total for different-sized slices of pie, pizza, or cake; different parts of a mile; or different colors of fraction manipulatives. This provides access because students can use a model or visual representation to solve and are choosing how to solve the problem. Ask students to develop the procedure based on prior knowledge: "Use what you know about fractions to figure out how to add the following fractions: . . ."

Figure B.1: Suggested answers for activity 1.4.   continued →

Task	Rationale	Ways to Adapt the Task to Increase the Cognitive Demand
Level 1: Angles Grade 7 Lines AB and CF are parallel. Name pairs of angles that are: 　a. Vertical angles 　b. Supplementary angles 　c. Alternate interior angles 　d. Corresponding angles	The task asks students to identify pairs of angles that meet specific definitions and vocabulary. Students must recall the properties of the different types of angles. If students did not already know the definitions and vocabulary, they would not be able to solve the task.	Allow students to discover angle relationships and justify their findings: If AB ∥ CF, determine which angles are congruent and justify your reasoning. This adaptation provides access as students are able to make sense of the relationships in many different ways. While a learner who struggles might not determine all of the relationships with more scaffolding, the learner may be able to determine some of them on his or her own.

APPENDIX C
Suggested Answers for Activity 3.2

Probing Students' Thinking	Exploring Mathematical Meanings and Relationships	Generating Discussion
• Clarifies student thinking • Enables students to elaborate their own thinking for their own benefit and for the class	• Points to underlying mathematical relationships and meanings • Makes links between mathematical ideas	• Enables other members of the class to contribute and comment on ideas under discussion
Why did you use that algorithm to solve the problem? How did you determine the scale for your graph? How did you come up with your answer? Explain to me how you got that product.	How does your table relate to your graph? Could you use that formula to find the volume of a different object? What does x represent in the equation? What is staying the same in the graph? Why is it staying the same? What is a word problem that could be modeled by this expression?	Who agrees with what Ryan said? Why do you agree? What did Keisha say? What else did you notice about the graph of the function? How could you describe Cameron's solution process in your own words?
Eliciting Procedures or Facts	**Inquiring About Other Mathematical Topics**	**Asking Nonmathematical Questions**
• Elicits a mathematical fact, procedure, or definition • Requests the result (only) of a mathematical procedure • Requests units or terminology • Requires a yes-or-no or single-word answer	• Relates to teaching and learning mathematics, but does not directly relate to the task or mathematical ideas of focus for the lesson	• Does not relate to teaching and learning mathematics • May relate to the context of the lesson or task
What is 6×7? What is the square root of 16? What is the definition of a quadrilateral? What is the denominator? Is the graph linear or quadratic?	Which problem was the most difficult for you to solve?	Who has ever been skiing?

Figure C.1: Suggested answers for activity 3.2.

APPENDIX D
Additional Support for Rating Tasks

This appendix provides additional support for rating tasks at each level. This information might be helpful for your collaborative team to develop consistency when rating tasks.

Level 1

Level 1 tasks involve memorization or recall. Tasks that focus on vocabulary, mathematical facts (such as timed tests), most fill-in-the-blank questions, or worksheets that require one-word (or one-number) responses are level 1. Note taking is also considered a level 1 task, as students' main mathematical activity is reproducing given mathematical facts, formulas, procedures, or rules.

When deciding between levels 1 and 2, consider whether the task requires students to use a procedure. Tasks in which students apply a memorized or rote procedure are often hard to classify, as they appear to meet the criteria for level 1 and level 2. If the task only requires students to retrieve the answer from memory, the task rates a level 1. If the task requires the student to apply a procedure, even if the procedure is rote or memorized, the task rates a level 2.

Level 2

Level 2 tasks focus on the correctness of steps in the procedure and on the correctness of the result, but not on making connections or understanding why the procedure works or makes sense to use in a given situation or context. Tasks that rate a level 2 are algorithmic, often having a single prescribed strategy or pathway to follow.

Tasks that require students to solve or practice several of the same type of problem, in which there are too many problems to spend time thinking about or creating meaning for mathematical ideas or processes, in which each problem follows the same template, or in which the directions indicate exactly how to solve each problem, also rate a level 2. With these tasks, there are no expectations for students to make sense of the procedure or underlying mathematical idea.

An important demarcation line exists between levels 1 and 2, and between levels 3 and 4. Levels 1 and 2 represent tasks with low cognitive demand or low-level tasks (Stein et al., 2009). Levels 3 and 4 represent tasks with high cognitive demand or high-level tasks. Again, this is not to suggest the levels 1 and 2 tasks are unnecessary or should never be used, but only to raise awareness that tasks at level 1 and level 2 provide students with opportunities to demonstrate their ability to memorize and perform rote procedures. If we want students to make sense of mathematics and engage in problem solving, we need to provide opportunities for thinking, reasoning, and problem solving in the tasks we use during mathematics instruction (Stein et al., 2009).

When deciding between levels 2 and 3, consider whether the task promotes a mathematical connection or understanding of a mathematical concept or procedure as well as providing multiple access points for students to engage in the task. Does the task help students make sense of some mathematics or mathematical procedure (level 3), or is the task providing an opportunity for students to practice or apply mathematics they already know (level 2)?

Level 3

Tasks at a level 3 allow students to develop an understanding of a mathematical concept, idea, or procedure, but do not explicitly require students to explain why, justify, compare, or reflect. Level 3 tasks often involve a variety of representations or strategies, thus allowing access for more students compared to tasks at level 1 or 2, as there are multiple points of entry. Teachers may prescribe a solution strategy, pathway, or procedure with the purpose of developing students' understanding of a larger mathematical concept, idea, or procedure. For example, a level 3 task may direct students to plot given points, with a prompt to add an additional point that follows the same pattern, for the purpose of promoting an understanding of a larger mathematical idea such as slope, translations, or inverse functions. Level 3 tasks can also provide opportunities for problem solving, but either the problem itself or the mathematics in the problem lacks the complexity to warrant deep reasoning or reflection on mathematical ideas. Problem-solving tasks in which students must develop their own solution strategy but are not asked to explain their thinking are also rated as a level 3.

When deciding between levels 3 and 4, consider whether the task contains an explicit prompt for an explanation, mathematical reasoning, proof, evidence, justification, or an argument. We often hear conversations about what "counts" as an explicit prompt for explanation or justification, or whether a prompt is "strong enough" to warrant a level 4. The prompt for a level 4 task can be as simple as *Explain* or *How do you know?* if the task itself (without the prompt) meets the criteria for a level 3. Consider whether an acceptable response to the prompt—not the best possible response nor the shortest possible response—would provide evidence of students' thinking, reasoning, and understanding. If so, then the prompt is considered to be "strong enough" to warrant a level 4.

Level 4

Level 4 tasks provide opportunities for problem solving or sense making in which students must also explain their thinking. This includes two parts: (1) complex, multistep problems for which students must develop their own solution strategies and illustrate or explain their reasoning about mathematics, and (2) tasks that provide opportunities to develop an understanding of mathematical concepts or procedures. As with a level 3 task, if the teacher prescribes a method or pathway, the purpose is to develop an understanding of a larger mathematical concept or procedure. Beyond a level 3 task, the level 4 task also has an explicit requirement to engage in activities such as explaining, justifying, comparing, reflecting on, or describing the mathematical connections or underlying ideas. Level 4 tasks may provide students with opportunities to use and discuss more than one strategy or representation simultaneously, where the student would need to use two or more representations to solve and explain the problem. Note that while asking students to identify patterns or express a pattern or relationship using a rule (verbal, symbolic, or other) provides the opportunity for students to make generalizations, in order to rate at level 4, the task would need to explicitly prompt students to explain, justify, or provide evidence (for example, a diagram) of why the pattern or generalization is true or works in all cases.

APPENDIX E
List of Figures and Videos

Table E.1 highlights the figures and videos throughout the book. It provides the appropriate grade level and mathematical topic as applicable.

Table E.1: Figures and Videos

Title	Grade Level and Mathematical Topic	Page Number
Introduction		
Figure I.1: The TQE process.		2
Figure I.2: Play button icon.		3
Figure I.3: Task icon.		3
Chapter 1		
Figure 1.1: The Leftover Pizza task (grade 6).	Grade 6: Fraction division	8
Figure 1.2: Benchmark Tasks grid.	Grade 4: Remainders in division Grade 6: Area of a trapezoid Grade 7: Integer operations	13
Figure 1.3: Benchmark Tasks recording sheet.		14
Figure 1.4: IQA Potential of the Task rubric.		15
Figure 1.5: Tasks for activity 1.3.	Grade 2: Patterning Grade 3: Comparing fractions Grade 5: Division story problems; properties of multiplication Grade 6: Measures of center, area Grade 7: Solving multistep equations Algebra: Quadratic functions	17
Figure 1.6: Tasks for activity 1.4.	Grade K: Number pairs that make 10 Grade 5: Adding fractions with unlike denominators Grade 7: Geometry (angle relationships)	22

continued →

Chapter 2		
Video: Leftover Pizza Lesson Version 1	Grade 6: Fraction division	30
Figure 2.1: Leftover Pizza lesson version 2.	Grade 6: Fraction division	31
Figure 2.2: IQA Implementation of the Task rubric.		33
Figure 2.3: Benchmark samples of implementation.	Grade 4: Fraction multiplication; polygons Grade 6: Fraction division Algebra: Simplifying expressions	35
Figure 2.4: Tasks for activity 2.3.	Grade 4: Remainders in division Grade 6: Area of a trapezoid Grade 7: Integer operations	39
Figure 2.5: Student work samples for 26 Divided by 4 task.	Grade 4: Remainders in division	40
Figure 2.6: Student work samples for Area of a Trapezoid task.	Grade 6: Area of a trapezoid	41
Figure 2.7: Student work samples for Rules for Integers task.	Grade 7: Integer operations	42
Figure 2.8: Characteristics of student work when cognitive demand is maintained, increased, or decreased.		44
Figure 2.9: Implementation Observation Tool.		47
Figure 2.10: The Father and Son Race task (algebra).	Algebra: System of linear equations	48
Video: Father and Son Race Lesson Version 1	Algebra: System of linear equations	49
Figure 2.11: Father and Son Race lesson version 2.	Algebra: System of linear equations	50
Chapter 3		
Video: Father and Son Race Lesson Version 1 (Revisit)	Algebra: System of linear equations	49
Figure 3.1: Father and Son Race lesson version 2 extension.	Algebra: System of linear equations	58
Figure 3.2: Framework for considering different types of questions.		60
Figure 3.3: Template and questions for activity 3.2.		63
Figure 3.4: Questions for 26 Divided by 4 and Area of a Trapezoid tasks.	Grade 4: Remainders in division Grade 6: Area of a trapezoid	65
Figure 3.5: IQA Teacher's Questions rubric.		67

List of Figures and Videos

Video: 26 Divided by 4 Lesson	Grade 4: Remainders in division	69
Figure 3.6: Examples of questions from the 26 Divided by 4 lesson.	Grade 4: Remainders in division	69
Chapter 4		
Video: Leftover Pizza Lesson Version 1 (Revisit)	Grade 6: Fraction division	30
Video: Change Unknown Lesson	Grade 1: Change unknown story problem	77
Figure 4.1: IQA Teacher's Linking rubric.		82
Figure 4.2: IQA Teacher's Press rubric.		83
Figure 4.3: Algebraic and Graphical Solutions to Equations task.	Algebra: Algebraic and graphical solutions to equations	84
Video: Algebraic and Graphical Solutions to Equations Lesson	Algebra: Algebraic and graphical solutions to equations	85
Chapter 5		
Video: Elapsed Time Lesson	Grade 3: Elapsed time on a number line	94
Video: Squares and Rectangles Lesson	Grade 3: Characteristics of squares and rectangles	96
Figure 5.1: IQA Students' Linking rubric.		100
Figure 5.2: IQA Students' Providing rubric.		100
Video: Decimals on a Number Line Lesson	Grade 4: Identifying tenths and hundredths on a number line	102
Chapter 6		
Video: Decimals on a Number Line Lesson (Revisit)	Grade 4: Identifying tenths and hundredths on a number line	102
Figure 6.1: IQA summary sheet.		108
Figure 6.2: IQA summary sheet for the Decimals on a Number Line lesson.		112
Figure 6.3: The Bridge Pattern task.	Grade 7: Patterning (linear relationships)	113
Video: Bridge Pattern Lesson	Grade 7: Patterning (linear relationships)	113
Figure 6.4: IQA summary sheet for the Bridge Pattern lesson.	Grade 7: Patterning (linear relationships)	117
Figure 6.5: Sample teacher's reflection on the Bridge Pattern lesson.	Grade 7: Patterning (linear relationships)	119

References and Resources

Boaler, J., & Brodie, K. (2004). The importance, nature, and impact of teacher questions. In D. E. McDougall & J. A. Ross (Eds.), *Proceedings of the twenty-sixth annual meeting of the North American Chapter of the International Group for the Psychology of Mathematics Education* (pp. 774–782). Toronto: Ontario Institute for Studies in Education, University of Toronto.

Boaler, J., & Staples, M. (2008). Creating mathematical futures through an equitable teaching approach: The case of Railside School. *Teachers College Record, 110*, 608–645.

Boston, M. D. (2012). Assessing the quality of mathematics instruction. *Elementary School Journal, 113*(1), 76–104.

Boston, M. D. (2017). *Instructional Quality Assessment Classroom Observation Tool: Rater packet* [Unpublished document].

Boston, M. D., & Candela, A. G. (2018). The Instructional Quality Assessment as a tool for reflecting on instructional practice. *ZDM: The International Journal on Mathematics Education, 50*(3), 427–444.

Boston, M. D., Henrick, E. C., Gibbons, L. K., Berebitsky, D., & Colby, G. T. (2016). Investigating how to support principals as instructional leaders in mathematics. *Journal of Research on Leadership Education, 12*(3), 183–214.

Boston, M. D., & Smith, M. S. (2009). Transforming secondary mathematics teaching: Increasing the cognitive demands of instructional tasks used in teachers' classrooms. *Journal for Research in Mathematics Education, 40*(2), 119–156.

Boston, M. D., & Smith, M. S. (2011). A "task-centric approach" to professional development: Enhancing and sustaining mathematics teachers' ability to implement cognitively challenging mathematical tasks. *ZDM: The International Journal on Mathematics Education, 43*(6–7), 965–977.

Boston, M. D., & Wilhelm, A. G. (2015). Middle school mathematics instruction in instructionally focused urban districts. *Urban Education, 52*(7), 829–861.

Cai, J., Moyer, J. C., Wang, N., Hwang, S., Nie, B., & Garber, T. (2013). Mathematical problem posing as a measure of curricular effect on students' learning. *Educational Studies in Mathematics, 83*(1), 57–69.

Candela, A. G. (2016). Using the Instructional Quality Assessment observation tool in a professional development capacity. In M. B. Wood, E. E. Turner, M. Civil, & J. A. Eli (Eds.), *Proceedings of the 38th annual meeting for the North American Chapter for the Psychology of Mathematics Education* (p. 418). Tucson: University of Arizona.

Carpenter, T. P., Fennema, E., Franke, M. L., Levi, L., & Empson, S. B. (2014). *Children's mathematics: Cognitively guided instruction* (2nd ed.). Portsmouth, NH: Heinemann.

Cohen, E. G., & Lotan, R. A. (2014). *Designing groupwork: Strategies for the heterogeneous classroom* (3rd ed.). New York: Teachers College Press.

Darling-Hammond, L. (2000). Teacher quality and student achievement. *Education Policy Analysis Archives, 8*(1). Accessed at https://epaa.asu.edu/ojs/article/view/392/515 on September 28, 2018.

Darling-Hammond, L. (2014/2015). Want to close the achievement gap?: Close the teaching gap. *American Educator, 38*(4), 14–18.

Dixon, J. K., Brooks, L. A., & Carli, M. R. (2018). *Making sense of mathematics for teaching the small group.* Bloomington, IN: Solution Tree Press.

Dixon, J. K., Nolan, E. C., & Adams, T. L. (2016). *What does it mean to teach mathematics with focus, coherence, and rigor, and how is it achieved?* [White paper]. Bloomington, IN: Solution Tree.

Dixon, J. K., Nolan, E. C., Adams, T. L., Brooks, L. A., & Howse, T. D. (2016). *Making sense of mathematics for teaching grades K–2.* Bloomington, IN: Solution Tree Press.

Dixon, J. K., Nolan, E. C., Adams, T. L., Tobias, J. M., & Barmoha, G. (2016). *Making sense of mathematics for teaching grades 3–5.* Bloomington, IN: Solution Tree Press.

Fennema, E., Carpenter, T. P., Franke, M. L., Levi, L., Jacobs, V. R., & Empson, S. B. (1996). A longitudinal study of learning to use children's thinking in mathematics instruction. *Journal for Research in Mathematics Education, 27*(4), 403–434.

Gokbel, E. N., & Boston, M. D. (2015). Considering students' responses in determining the quality of teachers' questions during mathematical discussions. In T. G. Bartell, K. N. Bieda, R. T. Putnam, K. Bradfield, & H. Dominguez (Eds.), *Proceedings of the 37th annual meeting of the North American Chapter of the International Group for the Psychology of Mathematics Education* (p. 1172). East Lansing: Michigan State University.

Grouws, D. A., Tarr, J. E., Chávez, Ó., Sears, R., Soria, V. M., & Taylan, R. D. (2013). Curriculum and implementation effects on high school students' mathematics learning from curricula representing subject-specific and integrated content organizations. *Journal for Research in Mathematics Education, 44*(2), 416–463.

Henningsen, M. A., & Stein, M. K. (1997). Mathematical tasks and student cognition: Classroom-based factors that support and inhibit high-level mathematical thinking and reasoning. *Journal for Research in Mathematics Education, 28*(5), 524–549.

Jackson, K. J., Shahan, E. C., Gibbons, L. K., & Cobb, P. A. (2012). Launching complex tasks. *Mathematics Teaching in the Middle School, 18*(1), 24–29.

Kazemi, E., & Hintz, A. (2014). *Intentional talk: How to structure and lead productive mathematical discussions.* Portsmouth, NH: Stenhouse.

Lappan, G., & Briars, D. (1995). How should mathematics be taught? In I. M. Carl (Ed.), *Seventy-five years of progress: Prospects for school mathematics* (pp. 131–156). Reston, VA: National Council of Teachers of Mathematics.

Manouchehri, A., & Lapp, D. A. (2003). Unveiling student understanding: The role of questioning in instruction. *Mathematics Teacher, 96*(8), 562–566.

McClure, L. (2011). *Using low threshold high ceiling tasks.* Cambridge, England: NRICH Project, University of Cambridge. Accessed at http://nrich.maths.org/7701 on September 28, 2018.

Mehan, H. (1979). *Learning lessons: Social organization in the classroom.* Cambridge, MA: Harvard University Press.

Merritt, E. G., Rimm-Kaufman, S. E., Berry, R. Q., III, Walkowiak, T. A., & McCracken, E. R. (2010). A reflection framework for teaching mathematics. *Teaching Children Mathematics, 17*(4), 238–248.

Metz, M. L. D. (2007). *A study of high school mathematics teachers' ability to identify and create questions that support students' understanding of mathematics* (Doctoral dissertation). University of Pittsburgh, Pittsburgh, PA.

Michaels, S., O'Connor, M. C., Hall, M. W., & Resnick, L. B. (2010). *Accountable Talk sourcebook: For classroom conversation that works.* Pittsburgh, PA: University of Pittsburgh Institute for Learning.

National Council of Teachers of Mathematics. (2000). *Principles and standards for school mathematics.* Reston, VA: Author.

National Council of Teachers of Mathematics. (2014). *Principles to actions: Ensuring mathematical success for all.* Reston, VA: Author.

National Governors Association Center for Best Practices & Council of Chief State School Officers. (2010). *Common Core State Standards for mathematics.* Washington, DC: Authors. Accessed at www.corestandards.org/assets/CCSSI_Math%20Standards.pdf on September 28, 2018.

National Research Council. (2001). *Adding it up: Helping children learn mathematics.* Washington, DC: The National Academies Press.

Nolan, E. C., Dixon, J. K., Roy, G. J., & Andreasen, J. B. (2016). *Making sense of mathematics for teaching grades 6–8.* Bloomington, IN: Solution Tree Press.

Nolan, E. C., Dixon, J. K., Safi, F., & Haciomeroglu, E. S. (2016). *Making sense of mathematics for teaching high school.* Bloomington, IN: Solution Tree Press.

Parke, C., Boston, M. D., Thomas-Browne, C., & Orsini, D. (in press). *Investigating the relationships between productive beliefs and practices in equitable and ambitious mathematics teaching* [Unpublished manuscript].

Resnick, L. B., Asterhan, C., & Clarke, S. (2015). *Socializing intelligence through academic talk and dialogue.* Washington, DC: American Educational Research Association.

Schoenfeld, A. H. (2002). Making mathematics work for all children: Issues of standards, testing, and equity. *Educational Researcher, 31*(1), 13–25.

Stein, M. K., & Lane, S. (1996). Instructional tasks and the development of student capacity to think and reason: An analysis of the relationship between teaching and learning in a reform mathematics project. *Educational Research and Evaluation, 2*(1), 50–80.

Stein, M. K., Smith, M. S., Henningsen, M. A., & Silver, E. A. (2009). *Implementing standards-based mathematics instruction: A casebook for professional development.* Reston, VA: National Council of Teachers of Mathematics.

Sztajn, P., Confrey, J., Wilson, P. H., & Edgington, C. (2012). Learning trajectory based instruction: Toward a theory of teaching. *Educational Researcher, 41*(5), 147–156.

Wagner, D., & Herbel-Eisenmann, B. (2009). Re-mythologizing mathematics through attention to classroom positioning. *Educational Studies in Mathematics, 72*(1), 1–15.

Weiss, I. R., Pasley, J. D., Smith, P. S., Banilower, E. R., & Heck, D. J. (2003, May). *Looking inside the classroom: A study of K–12 mathematics and science education in the United States.* Chapel Hill, NC: Horizon Research. Accessed at www.horizon-research.com/horizonresearchwp/wp-content/uploads/2013/04/highlights.pdf on September 28, 2018.

Index

NUMBERS

26 Divided by 4 lesson
    considering different types of tasks (introductory activity) and, 12–13
    creating questions (introductory activity) and, 63–65
    identifying evidence of students' thinking and reasoning in students' work (application activity) and, 39, 40, 43
    rating the teacher's questions in the 26 Divided by 4 lesson (application activity) and, 68–70

A

academic tasks, 3
access to the problem
    context and, 64
    discussions and, 79
    IQA rubrics and, 123
    low threshold, high ceiling tasks and, 10
    prior knowledge and, 14
    rating implementation of the task—Father and Son Race lesson (application activity) and, 51
    teacher's questions and, 66, 70
accurate mathematics, press for, 82
activities (mathematical tasks), 5
Adding Fractions with Unlike Denominators task, 22, 23, 24, 137
Addition and Subtraction of Integers task, 11, 12–13
administrators, 3, 122
Algebraic and Graphical Solutions lesson
    rating teacher's linking and teacher's press (application activity) and, 83–87
    teacher's linking and, 97–98
    teacher's press and, 89
algorithms, 11, 19–20

Angles task, 22, 24, 138
application activities
    considering the teacher's reflection and next steps, 119–121
    determining when it is appropriate to ask a follow-up question, 87–89
    identifying evidence of students' thinking and reasoning in students' work, 38–44
    moving from tasks to implementation, 45
    rating implementation of the task—Father and Son Race lesson, 48–50, 50–52
    rating mathematical tasks using the Potential of the Task rubric, 16–20
    rating students' linking and students' providing, 101–104
    rating teacher's linking and teacher's press, 83–87
    rating the teacher's questions in the 26 Divided by 4 lesson, 68–70
    using all IQA rubrics, 121–123
    using the Implementation of the Task rubric, 34–38
    using the IQA Potential of the Task rubric to rate and adapt tasks, 21–24
Area of a Trapezoid task
    creating questions (introductory activity) and, 63–65
    identifying evidence of students' thinking and reasoning in students' work (application activity) and, 39, 41, 43
asking nonmathematical questions (question types). *See also* teacher's questions
    creating questions (introductory activity) and, 64–65
    framework for different types of questions and, 60
    IQA Teacher's Questions rubric and, 67
    rating the teacher's questions in the 26 Divided by 4 lesson (application activity) and, 68–70

and suggested answers, 139
assessments
    teacher assessment and self-assessment, 125–126
    teacher's questions and, 70, 71, 80
    TQE process and, 2, 91, 95

B

Boaler, J., 59
Bridge Pattern lesson
    considering the teacher's reflection and next steps (application activity) and, 119–121
    rating a whole-class lesson (introductory activity) and, 112–117
    sample teacher's reflection on, 118–119
broad questions, 66. See also teacher's questions
Brodie, K., 59

C

challenging, definition of, 14
Change Unknown lesson
    following up on students' contributions—teacher's press (introductory activity), 76–79
    linking and, 81
classroom norms, 75, 80
classroom observations, 46–47
clocks, 94
cognitive demand
    changing levels of, 43–44, 52
    considerations when rating tasks and, 25
    increasing levels of, 21, 23–24, 66
    IQA Implementation Observation Tool and, 46–47
    IQA Implementation of the Task rubric and, 32, 34
    IQA Potential of the Task rubric and, 14–16
    and learners who struggle in mathematics, 123
    and rating levels, 141
    students' work as evidence for, 44
cognitive potential, importance of, 7
cognitively challenging tasks
    characteristics of work for, 44
    difficulty and, 14
    IQA Potential of the Task rubric and, 123
    prior knowledge and, 25
    research on, 45
combining like terms, 37
Common Core standards and learning progression, 26
Common Core State Standards (CCSS) for mathematics, 9
comparing two mathematics lessons (introductory activities), 29–32
composing numbers, 21
computations
    IQA Implementation of the Task rubric and, 37
    IQA Potential of the Task rubric and, 19–20
    Leftover Pizza lesson and, 32
    potential of the task and, 11
conceptual connections, 115
conceptual tasks, 23
conceptual understanding, 26, 81–82, 101
conjectures, forming, 19
connections and mathematical thinking, 9
connections between representations, 18, 19
considerations when rating tasks
    about, 24–25
    aligning with learning goals and, 26–27
    defining the task and, 25
    implications of higher-level thinking and, 25–26
considering different types of tasks (introductory activity), 10–14
considering the teacher's reflection and next steps (application activity), 119–121
context
    increasing cognitive demands and, 23
    IQA Potential of the Task rubric and, 18, 49
    Leftover Pizza lesson and, 8–10, 32
    potential of the task and, 18–20
    procedures and, 27
    teacher's questions and, 61, 64, 70
    word problems and story problems and, 20
creating questions (introductory activity), 63–65
curricular materials, rating tasks in, 25

D

data, 19, 125
Decimals on a Number Line lesson
    IQA summary sheet for, 112
    rating a small-group lesson (introductory activity) and, 107–112
    rating students' linking and students' providing (application activities) and, 101–104
decomposing numbers, 21
demanding, definition of, 14
demands of the task, 52. *See also* cognitive demand
determining when it is appropriate to ask a follow-up question (application activity), 87–89
diagrams, 9, 37
discourse actions. *See also* teacher's linking; teacher's press; teacher's questions
    about, 55
    evidence of students' thinking and, 91, 116
    explanations and, 94
    feedback and, 120
    following up on students' contributions—teacher's linking (introductory activity) and, 74–75
    following up on students' contributions—teacher's press (introductory activity) and, 77–78
    frequency of, 99
    IQA rubrics and, 122
    IQA Teacher's Linking and Teacher's Press rubric, 82
    rating a small-group lesson (introductory activity) and, 111
    student confidence and status and, 123
    students' contribution and, 97
    teacher's linking and teacher's press and, 101
    teacher's questions and follow up (introductory activity) and, 79–80
discourse pattern, Initiate-Response-Evaluate (IRE), 78
discussions. *See* small-group discussion; students' contribution; whole-class discussion
distributive property, 37
division. *See also* 26 Divided by 4 lesson; fraction division; Leftover Pizza lesson
    Division Story Problems task, 17, 20
    Division with Remainders task, 11
drawing, 8

E

Elapsed Time lesson, 93–95
elementary schools and high-level instructional tasks, 7
eliciting procedures or facts (question types). *See also* teacher's questions
    cognitive demands and, 66
    creating questions (introductory activity) and, 63–64
    framework for different types of questions and, 60
    IQA Teacher's Press rubric and, 67
    rating the teacher's questions in the 26 Divided by 4 lesson (application activity) and, 68–70
    sorting questions (introductory activity) and, 61–63
    suggested answers for, 139
engagement
    IQA Implementation of the Task rubric and, 32, 43, 109
    linking and, 123
    maintaining the level of, 70
    rating levels and, 142
entry points. *See* access to the problem
equations, Multistep Equations task, 17, 20
evidence
    of knowledge, 98
    of mathematical work and thinking, 91
    of students' mathematical thinking and understanding, 95
    of students' thinking and reasoning, 81
    students' work and cognitive demands and, 44
    TQE process and, 2, 91
    of understanding, 36
evidence of students' thinking and reasoning in students' work, identifying (application activity)
    discuss, 43–44
    engage, 38
    tasks for, 39–41
examining students' contributions—students' linking (introductory activity), 93–95
examining students' contributions—students' providing (introductory activities), 95–97
examples and exercises (mathematical tasks), 5
explain and justify solutions (classroom norm), 75, 80

explanations. *See also* justifications
- conceptual understanding and, 101
- and evidence of students' mathematical thinking and understanding, 95
- following up on students' contributions—teacher's press (introductory activity) and, 77–79
- gaps in understanding and, 14
- identifying different types of questions (introductory activity) and, 59
- IQA Implementation of the Task rubric and, 36, 49
- IQA Potential of the Task rubric and, 18, 19
- Leftover Pizza lesson and, 31, 32
- potential of the task and, 11
- procedures and, 26, 44
- rating levels and, 142
- students' providing and, 96, 103
- teacher's press and, 78, 86

exploring mathematical meaning and relationships (question types). *See also* teacher's questions
- creating questions (introductory activity) and, 63–64
- how teachers' questions impact implementation (introductory activity) and, 65–67
- identifying different types of questions (introductory activity) and, 59
- IQA Teacher's Questions rubric and, 67–68
- rating a small-group lesson (introductory activity) and, 109, 110
- rating the teacher's questions in the 26 Divided by 4 lesson (application activity) and, 68–70
- sorting questions (introductory activity) and, 61–63
- suggested answers for, 139

F

facts, teacher's press and, 82

Father and Son Race lesson
- determining when it is appropriate to ask a follow-up question (application activity) and, 87–88
- identifying different types of questions (introductory activity) and, 57–61
- Initiate-Response-Evaluate (IRE) discourse pattern and, 80
- questions related to context and, 64
- rating implementation of the task—Father and Son Race lesson (application activity), 48–50
- rating implementation of the task—Father and Son Race lesson version 2 (application activity), 50–52
- teacher questions and, 65–66
- teacher's linking and teacher's press and, 98–99

feedback
- discourse actions and, 120
- IQA rubrics and, 3
- language of rubrics and, 126
- rating a small-group lesson (introductory activity) and, 111–112

fill-in-the-blank questions, 141

following up on students' contributions—teacher's linking (introductory activity), 73–76

following up on students' contributions—teacher's press (introductory activity), 76–79

formative assessments
- hearing students' thinking and, 80
- teacher's questions, 71
- TQE process and, 2, 91, 95

fraction division, 8, 9, 31–32. *See also* division; Leftover Pizza lesson

Fraction Pizza task, 18, 19

fraction tiles, 10

fractions, Adding Fractions with Unlike Denominators task, 22, 23, 24, 137

Fractions Operation Model task, 35, 37

G

generalizations, 44, 114, 142

generating discussion (question types). *See also* teacher's questions
- creating questions (introductory activity) and, 63–64
- how teachers' questions impact implementation (introductory activity) and, 65–67
- identifying different types of questions (introductory activity) and, 59
- IQA Teacher's Questions rubric and, 67–68

rating a small-group lesson (introductory activity) and, 109, 110
rating the teacher's questions in the 26 Divided by 4 lesson (application activity) and, 68–70
sorting questions (introductory activity) and, 61–63
suggested answers for, 139
goals for students' learning, task levels and, 26
grade 1, 37, 76
grade 2, 18, 19, 26
grade 3, 18, 93, 95
grade 4, 11, 26, 63, 101
grade 5, 17, 18, 24, 26
grade 6, 8, 11, 18, 29, 63
grade 7, 11, 17, 113
graphs, 18, 50, 51, 88
group discussion. *See* small-group discussion; whole-class discussion

H

high ceiling and low threshold tasks, 10
high schools and high-level instructional tasks, 7
higher-level thinking, considering implications of, 25–26
high-level tasks, 141
holistic ratings, 33, 46, 109
homework sets (mathematical tasks), 5
how teachers' questions impact implementation (introductory activity), 65–67

I

identifying different types of questions (introductory activity), 57–61
identifying evidence of students' thinking and reasoning in students' work (application activity)
    discuss, 43–44
    engage, 38
    tasks for, 39–41
implementation of the task. *See also* IQA Implementation Observation Tool; IQA Implementation of the Task rubric
    about, 29
    application activities, 34–45, 48–52

considerations for implementation based on classroom observation or videos, 46–47
introductory activities, 29–32
summary, 52
transition activity—how teachers' questions impact implementation, 52–53
Initiate-Response-Evaluate (IRE) discourse pattern, 78, 80, 89
inquiring about other mathematical topics (question types). *See also* teacher's questions
    creating questions (introductory activity) and, 64–65
    framework for different types of questions and, 60
    IQA Teacher's Questions rubric and, 67
    rating the teacher's questions in the 26 Divided by 4 lesson (application activity) and, 68–70
    suggested answers for, 139
instructional leaders, 125
instructional practices
    about using the IQA Toolkit as a tool to assess and improve, 107
    framework for, 125
    IQA rubrics as tools to reflect on and improve, 118–119
    rubrics and, 125
instructional quality, 67, 93, 119
Instructional Quality Assessment (IQA) Mathematics Toolkit, 2
integers, Addition and Subtraction of Integers task, 11, 12–13
interpretation, 9
introductory activities
    comparing two mathematics lessons, 29–32
    considering different types of tasks, 10–14
    creating questions, 63–65
    examining students' contributions—students' linking, 93–95
    examining students' contributions—students' providing, 95–97
    following up on students' contributions—teacher's linking, 73–76
    following up on students' contributions—teacher's press, 76–79
    how teachers' questions impact implementation, 65–67

identifying different types of questions, 57–61
rating a small-group lesson, 107–112
rating a whole-class lesson, 112–117
solving a task, 7–10
sorting questions, 61–63
teacher's linking, teacher's press, and students' contributions, 97–100
teacher's question and follow up, 79–80
investigation and mathematical properties, 27
IQA Implementation Observation Tool. *See also* implementation of the task
    about, 46
    how teachers' questions impact implementation (introductory activity) and, 65–67
    rating a small-group lesson (introductory activity) and, 108, 109
    rating a whole-class lesson (introductory activity) and, 114
    rating scale, 47
    teacher's linking, teacher's press, and students' contributions (introductory activity) and, 97
    teacher's questions and follow up (introductory activity) and, 79–80
    using all IQA rubrics (application activity) and, 122
IQA Implementation of the Task rubric. *See also* implementation of the task
    about, 32–34
    Bridge Pattern lesson and, 120–121
    examining students' contributions—students' linking (introductory activity) and, 94
    how teachers' questions impact implementation (introductory activity) and, 65–67
    identifying evidence of students' thinking and reasoning in students' work (application activity) and, 38, 43–44
    IQA Implementation Observation Tool and, 47
    moving from tasks to implementation (application activity) and, 45
    rating a small-group lesson (introductory activity) and, 108, 109
    rating a whole-class lesson (introductory activity) and, 114–115
    rating implementation of the task—Father and Son Race lesson (application activity) and, 48–50
    rating implementation of the task—Father and Son Race lesson version 2 (application activity) and, 50–52
    rating scale, 33–34
    students' providing and, 104
    teacher's linking, teacher's press, and students' contributions (introductory activity) and, 97
    teacher's question and follow up (introductory activity) and, 79–80
    teacher's questions and, 67
    using all IQA rubrics (application activity) and, 121, 122
    using the Implementation of the Task rubric (application activity) and, 34–38
IQA Potential of the Task rubric. *See also* potential of the task
    about, 14–16
    access to the problem and, 123
    Bridge Pattern lesson and, 120
    examining students' contributions—students' linking (introductory activity) and, 94
    how teachers' questions impact implementation (introductory activity) and, 65–67
    identifying evidence of students' thinking and reasoning in students' work (application activity) and, 38, 43–44
    IQA Implementation of the Task rubric and, 32–33
    Leftover Pizza lesson and, 29–32
    moving from tasks to implementation (application activity) and, 45
    rating a small-group lesson (introductory activity) and, 107
    rating a whole-class lesson (introductory activity) and, 114
    rating implementation of the task—Father and Son Race lesson (application activity) and, 48–50
    rating scale, 15
    rating students' linking and students' providing (application activities) and, 102, 104
    rating teacher's linking and teacher's press (application activity) and, 83, 85

teacher's linking, teacher's press, and students' contributions (introductory activity) and, 97
teacher's questions and follow up (introductory activity) and, 79–80
using all IQA rubrics (application activity) and, 121, 122
using the Implementation of the Task rubric (application activity) and, 34–38
using the IQA Potential of the Task rubric to rate and adapt tasks (application activity) and, 21–24

IQA rubrics. *See also specific rubrics*
    about, 2
    data and, 125
    feedback and, 3
    implementation of, 125–126
    language of, 122
    professional development and, 45
    rating a small-group lesson (introductory activity) and, 111–112
    and tools to reflect on and improve instruction, 118–119

IQA Students' Linking rubric. *See also IQA Teacher's Linking rubric; students' linking; teacher's linking*
    about, 91, 100–101
    Decimals on a Number Line lesson and, 108
    rating a whole-class lesson (introductory activity) and, 115–116
    rating scale, 100
    rating students' linking and students' providing (application activities) and, 101–104
    using all IQA rubrics (application activity) and, 122

IQA Students' Providing rubric. *See also students' providing*
    about, 91, 100–101
    Decimals on a Number Line lesson and, 108
    rating a whole-class lesson (introductory activity) and, 117
    rating scale, 100
    rating students' linking and students' providing (application activities) and, 101–104
    using all IQA rubrics (application activity) and, 122

IQA Teacher's Linking rubric, 55. *See also IQA Students' Linking rubric; students' linking; teacher's linking*
    about, 80–81
    access to the problem and, 123
    discourse actions and, 82
    IQA Teacher's Questions rubric and, 110, 121
    rating a small-group lesson (introductory activity) and, 108, 110–111
    rating a whole-class lesson (introductory activity) and, 115
    rating scale, 82
    rating teacher's linking and teacher's press (application activity) and, 85–86
    using all IQA rubrics (application activity) and, 121, 122

IQA Teacher's Press rubric, 55. *See also teacher's press*
    about, 67–68, 80, 81–82
    access to the problem and, 123
    discourse actions and, 82
    IQA Teacher's Questions rubric and, 110, 121
    rating a small-group lesson (introductory activity) and, 108, 111
    rating a whole-class lesson (introductory activity) and, 116
    rating scale, 67, 83
    rating teacher's linking and teacher's press (application activity) and, 86–87
    using all IQA rubrics (application activity) and, 121, 122

IQA Teacher's Questions rubric
    about, 55
    access to the problem and, 123
    framework for, 60
    IQA Teacher's Linking rubric and Teacher's Press rubric and, 110, 121
    rating a small-group lesson (introductory activity) and, 108, 109–110
    rating a whole-class lesson (introductory activity) and, 115
    rating the teacher's questions in the 26 Divided by 4 lesson (application activity) and, 69–70
    students' providing and, 104
    teacher's linking, teacher's press, and students' contributions (introductory activity) and, 97

teacher's questions and follow up (introductory activity) and, 79–80
using all IQA rubrics (application activity) and, 121, 122

IQA Toolkit
  about, 107
  application activities, 119–123
  and improving instruction, 118–119
  introductory activities, 107–117
  IQA Potential of the Task rubric levels and, 16
  summary, 123
  teacher's linking and, 75
  teacher's press and, 77

J

justifications. *See also* explanations
  conceptual understanding and, 101
  following up on students' contributions—teacher's press (introductory activity) and, 77–79
  IQA Implementation of the Task rubric and, 114
  IQA Potential of the Task rubric and, 18
  potential of the task and, 11
  and rating levels, 142
  students' providing and, 96, 103
  teacher's press and, 78
  understanding and, 14, 95
just-in-time scaffolding, 66, 71

K

kindergarten, 21, 37

L

language of rubrics, 122, 126
learner goals, aligning with, 26–27
learners who struggle in mathematics, 27, 123
learning progression, aligning with, 26
Leftover Pizza lesson
  comparing two mathematics lessons (introductory activities) and, 29–32
  following up on students' contributions—teacher's linking (introductory activity) and, 73–76
  Initiate-Response-Evaluate (IRE) discourse pattern and, 80
  IQA Implementation of the Task rubric and, 34
  linking and, 81
  potential of the task and, 8–10
  press and, 81–82
  teacher questions and, 66
  teacher's press and, 88–89
  using the Implementation of the Task rubric (application activity) and, 35, 36
lessons, assessing lessons with IQA rubrics, 125
level of actual thinking and reasoning, 33. *See also* IQA Implementation of the Task rubric
level of potential opportunities for thinking and reasoning, 33. *See also* IQA Potential of the Task rubric
levels. *See also* specific rubrics
  IQA Toolkit and, 16
  rating levels 1-4, 141–142
  rating tasks in curricular materials and other resources, 25
linear relationships, 49
linking. *See* IQA Students' Linking rubric; IQA Teacher's Linking rubric; students' linking; teacher's linking
low threshold and high ceiling tasks, 10
low-level tasks and rating levels, 141

M

make sense of each other's solutions (classroom norm), 75, 80
making sense. *See* sense making
Making Sense of Mathematics for Teaching Grades 3–5 (Dixon, Nolan, Adams, Tobias, and Barmoha), 23
Making Sense of Mathematics for Teaching series, 27, 75, 80
manipulatives, 8, 9, 10, 95
mastery/practice and IQA Potential of the Task rubric, 23
mathematical concepts, 85, 142
mathematical connections, 11, 37, 51, 142
mathematical facts, 59, 141
mathematical justifications and explanations, 91. *See also* explanations; justifications

Index

mathematical knowledge, teacher press and, 82, 87
Mathematical Practice 1, 9–10
Mathematical Practice 4, 10
Mathematical Practices, 9
mathematical problems and defining the task, 25
mathematical tasks, 5, 29. *See also* tasks
Mathematical Tasks Framework and Levels of Cognitive Demand, 2
mathematical thinking, 9, 59
mathematical understanding, 32
meaning, 18, 19, 21, 37, 49
memorization, 11, 20, 24, 66. *See also* recall
Metz, M., 59
middle schools and high-level instructional tasks, 7
misconceptions
    Father and Son Race lesson and, 88, 99
    fraction division and, 8, 31
    graphs and, 51
    instructional quality and, 122
    teacher questions and, 58, 66, 70
mixed-ability grouping, 80
models/modeling
    access to the problem and, 10
    direct modeling, 43
    fraction division and, 8, 9
    IQA Implementation Observation Tool and, 109
    IQA Implementation of the Task rubric and, 37
    IQA Potential of the Task rubric and, 19
    Leftover Pizza lesson and, 32
    procedures and memorization and, 44
moving from tasks to implementation (introductory activity), 45
moving from tasks to implementation (transition activity), 27–28
multiplication, 18, 20, 26
Multistep Equations task, 17, 20

N

National Council of Teachers of Mathematics (NCTM), 1, 9
national standards, 9
non-algorithmic thinking, 11
note taking, 11, 38, 43

number lines, 94
Number Pairs that Make 10, 21, 22, 23, 137

O

observations, 3, 46–47, 68

P

pattern blocks, 8, 10, 31
pattern identification, 19, 114
peers
    connecting to the work and thinking of, 91, 94–95, 99, 111
    mathematical communication and, 31, 50, 114
    mathematical thinking and, 75
pictures, 36–37
play button icon, about, 3
point of intersection, 49, 51
pointed questions, 66. *See also* teacher's questions
potential of the task. *See also* IQA Potential of the Task rubric
    about, 7
    application activities, 16–24
    considerations when rating tasks, 24–27
    introductory activities, 7–14
    and promoting learning, 52
    summary, 27
    transition activity—moving from tasks to implementation, 27–28
practice/mastery and IQA Potential of the Task rubric, 23
press. *See* IQA Teacher's Press rubric; teacher's press
Principles to Actions: Ensuring Mathematical Success for All (NCTM), 1, 9
prior knowledge, 14, 25, 47, 50
probing students' thinking (question types). *See also* teacher's questions
    creating questions (introductory activity) and, 63–64
    how teachers' questions impact implementation (introductory activity) and, 65–67
    identifying different types of questions (introductory activity) and, 59
    IQA Teacher's Questions rubric and, 67–68

rating a small-group lesson (introductory activity) and, 109, 110
rating the teacher's questions in the 26 Divided by 4 lesson (application activity) and, 68–70
sorting questions (introductory activity) and, 61–63
suggested answers for, 139
problem solving
  examining students' contributions—students' linking (introductory activity) and, 94
  IQA Potential of the Task rubric and, 18, 19
  potential of the task and, 11
  rating levels and, 141, 142
problems (mathematical tasks), 5
procedural or factual question types. *See* eliciting procedures or facts (question types)
procedures/procedural tasks
  exploring mathematical meaning and relationships (question types) and, 66
  fraction division and, 8, 9, 31–32
  increasing cognitive demands and, 24
  IQA Implementation of the Task rubric and, 37, 51
  IQA Potential of the Task rubric and, 19–20, 23
  learning progression and, 26–27
  potential of the task and, 11
  rating levels and, 141
  rote procedure, 141
  teacher's press and, 82
Process Standards of NCTM, 9
professional development, 45, 125
professional learning experiences, 125
progression of learning, aligning with, 26
prompts and rating levels, 142. *See also* teacher's linking; teacher's press; teacher's questions
proofs, 44
properties of multiplication, 18, 20
providing. *See* students' providing

Q

QR codes, about, 3
quadratic functions, 18
questions. *See* teacher's questions

R

rate of change, 49
rating a small-group lesson (introductory activity), 107–112
rating a whole-class lesson (introductory activity), 112–117
rating mathematical tasks using the Potential of the Task rubric (application activity), 16–20
rating students' linking and students' providing (application activities), 101–104
rating teacher's linking and teacher's press (application activity), 83–87
rating the teacher's questions in the 26 Divided by 4 lesson (application activity), 68–70
real-life applications, 37
reasoning, 1, 11, 26. *See also* explanations; justifications
recall
  IQA Potential of the Task rubric and, 20, 24
  potential of the task and, 11
  questions and, 59, 70
representations
  increasing cognitive demands and, 23
  IQA Potential of the Task rubric and, 18, 19, 85, 114
  mathematical thinking and, 9
research
  on cognitively challenging tasks, 45
  on decline of cognitive demands, 43
  language of rubrics and, 126
  on levels and type of thinking, 11
  on levels of tasks, 7
revoicing. *See also* discourse actions
  IQA Teacher's Linking rubric and, 115
  linking and, 81
  rating teacher's linking and teacher's press (application activity) and, 86
  teacher's question and follow up (introductory activity) and, 80
Road Sign Polygons task, 36, 37–38
rubrics, language of, 122, 126
Rules for Integers task, 39, 42, 43

S

say when you don't understand or when you don't agree (classroom norm), 75, 80
scaffolding, 51, 66, 71, 80
Science Quiz task, 17, 19
sense making
    explanations and, 26
    fraction division and, 9
    IQA Implementation of the Task rubric and, 37
    IQA Potential of the Task rubric and, 18, 19, 49, 114
    potential of the task and, 11
    and rating levels, 141, 142
sets of problems, 9
shape recognition, 37
Shapes Pattern task, 18, 19
"show your work," 20
Simplifying Expressions task (algebra), 35, 37
small groups
    IQA Implementation of the Task rubric and, 50
    rating a small-group lesson (introductory activity) and, 107–112
small-group discussion. *See also* whole-class discussion
    broad questions and, 66
    determining when it is appropriate to ask a follow-up question (application activity) and, 87–89
    evidence of students' mathematical work and thinking and, 91
    following up on students' contributions—teacher's linking (introductory activity) and, 74–75
    IQA Students' Linking and Students' Providing rubrics and, 100
    IQA Teacher's Linking and Teacher's Press rubric and, 82
    rating students' linking and students' providing (application activities) and, 102–103
    students' contribution and, 89
    students' providing explanations and justifications and, 103–104
    teacher's question and follow up (introductory activity) and, 80
solution methods, multiple, 10
solving a task (introductory activity), 7–10

sorting questions (introductory activity), 61–63
special education, 27
Squares and Rectangles lesson, 95–97
Standards for Mathematical Practice, 9–10
story problems, 20, 24, 37
strategies, 85, 94
structures, recognizing and using, 19
student achievement, 1, 123
student discourse, 80
students' contribution. *See also* small-group discussion; whole-class discussion
    discourse actions and, 97
    evidence in small-group and whole-class discussions, 91
    examining students' contributions—students' linking (introductory activity) and, 93–95
    examining students' contributions—students' providing (introductory activities) and, 95–97
    following up on students' contributions—teacher's linking (introductory activity) and, 73–76
    following up on students' contributions—teacher's press (introductory activity) and, 76–79
    IQA Students' Linking and Students' Providing rubrics and, 100
    linking and, 75, 78, 95
    teacher's linking and teacher's press and, 73, 89
    teacher's linking, teacher's press, and students' contribution (transition activity) and, 90
    teacher's linking, teacher's press, and students' contributions (introductory activity) and, 97–100
    teacher's question and follow up (introductory activity) and, 79–80
students' linking. *See also* IQA Students' Linking rubric; IQA Teacher's Linking rubric; teacher's linking
    about students' linking and student's providing, 93
    application activities, 101–104
    introductory activities, 93–100
    summary, 104
    teacher's linking and, 96
    teacher's linking and teacher's press and, 98
    transition activity—using all IQA rubrics, 105
students' providing. *See also* IQA Students' Providing rubric
    about students' linking and student's providing, 93

application activities, 101–104
introductory activities, 93–100
summary, 104
teacher's linking and teacher's press and, 98
transition activity—using all IQA rubrics, 105
students' thinking and reasoning, 51
students' thinking, hearing, 80
student-to-student discourse, 78
Subtraction of Integers task, Addition and, 11, 12–13
Swimming Pool Deck task, 18, 19

T

taking notes, 11, 38, 43
task icon, about, 3
tasks
    cognitive potential of, 7
    defining the task, 25
    and goals for students' learning, 26
    levels of, 24
    rating tasks in curricular materials and other resources, 25
    and supporting learning, 32
    TQE process and, 2, 5
tasks, considerations when rating. *See* considerations when rating tasks
tasks, implementation of. *See* implementation of the task
tasks, potential of the. *See* potential of the task
teacher collaboration, 125
teacher teams, 2
teachers, assessments and, 125–126
teacher's linking. *See also* discourse actions; IQA Students' Linking rubric; IQA Teacher's Linking rubric; students' linking
    about teacher's linking and teacher's press, 73
    application activities, 83–89
    discourse actions and, 101
    introductory activities, 73–80
    students' linking and, 96
    students' linking and students' providing and, 98
    teacher's linking, teacher's press, and students' contributions (introductory activity) and, 97–100

    teacher's questions and, 109
    transition activity— teacher's linking, teacher's press, and students' contribution, 90
teacher's press. *See also* discourse actions; IQA Teacher's Press rubric
    about teacher's linking and teacher's press, 73
    application activities, 83–89
    discourse actions and, 101, 111
    introductory activities, 73–80
    IQA Implementation Observation Tool and, 109
    students' linking and students' providing and, 98
    teacher's linking, teacher's press, and students' contributions (introductory activity) and, 97–100
    teacher's questions and, 109
    transition activity— teacher's linking, teacher's press, and students' contribution, 90
teacher's question and follow up (introductory activity), 79–80
teacher's questions. *See also* IQA Teacher's Questions rubric
    about, 57
    application activities, 68–70
    discourse actions and, 55
    formative assessments process and, 71
    how teachers' questions impact implementation (transition activity) and, 52–53
    introductory activities, 57–67
    IQA Implementation of the Task rubric and, 67
    IQA Teacher's Press rubric, 67–68
    question types, 61, 70
    summary, 70–71
    teacher's linking and teacher's press and, 109
    teacher's question and follow up (introductory activity) and, 79–80
    TQE process and, 2, 55
    transition activity—teacher's questions and follow-up contribution, 71
teacher-student discourse, 78
teaching, observing teaching, 3
thinking, considering implications of higher-level thinking, 25–26
thinking and reasoning, 1, 5, 9. *See also* evidence
TQE process
    about, 2

evidence of students' mathematical work and thinking and, 91
formative assessments and, 95
questions and, 55
tasks and task implementation and, 5

transition activities
how teachers' questions impact implementation, 52–53
how teachers' questions impact implementation, revisiting, 65–67
moving from tasks to implementation, 27–28
moving from tasks to implementation, revisiting, 45
teacher's linking, teacher's press, and students' contribution, 90
teacher's linking, teacher's press, and students' contributions, revisiting, 97–100
teacher's questions and follow-up, 71
teacher's questions and follow-up, revisiting, 79–80
using all IQA rubrics, 105
using all IQA rubrics, revisiting, 121–123

U

using all IQA rubrics (application activity), 121–123
using all IQA rubrics (transition activity), 105
using the IQA Potential of the Task rubric to rate and adapt tasks (application activity), 21–24

V

validation, teacher's press and, 78
videos/video clips, 3, 47, 68
vocabulary, rating levels and, 141

W

Water Fountain task (algebra), 17, 18
whiteboards, 31
whole-class discussion. *See also* small-group discussion
determining when it is appropriate to ask a follow-up question (application activity) and, 87–89
evidence of students' mathematical work and thinking and, 91
following up on students' contributions—teacher's linking (introductory activity) and, 74–75
IQA Implementation of the Task rubric and, 50
IQA Students' Linking and Students' Providing rubrics and, 100
IQA Teacher's Linking and Teacher's Press rubric and, 82
pointed questions and, 66
students' contribution and, 89
teacher's questions and follow up (introductory activity) and, 80
whole-class lesson, rating a (introductory activity), 112–117
whole-group instruction, 43
word problems, 20, 24, 37
worksheets and rating levels, 141

Y

y-intercept, 49

Making Sense of Mathematics for Teaching Grades K–2
Juli K. Dixon, Edward C. Nolan, Thomasenia Lott Adams, Lisa A. Brooks, and Tashana D. Howse
Develop a deep understanding of mathematics. With this user-friendly resource, grades K–2 teachers will explore strategies and techniques to effectively learn and teach significant mathematics concepts and provide all students with the precise, accurate information they need to achieve academic success.
BKF695

Making Sense of Mathematics for Teaching Grades 3–5
Juli K. Dixon, Edward C. Nolan, Thomasenia Lott Adams, Jennifer M. Tobias, and Guy Barmoha
Develop a deep understanding of mathematics. With this user-friendly resource, grades 3–5 teachers will explore strategies and techniques to effectively learn and teach significant mathematics concepts and provide all students with the precise, accurate information they need to achieve academic success.
BKF696

Making Sense of Mathematics for Teaching Grades 6–8
Edward C. Nolan, Juli K. Dixon, George J. Roy, and Janet B. Andreasen
Develop a deep understanding of mathematics. With this user-friendly resource, grades 6–8 teachers will explore strategies and techniques to effectively learn and teach significant mathematics concepts and provide all students with the precise, accurate information they need to achieve academic success.
BKF697

Making Sense of Mathematics for Teaching High School
Edward C. Nolan, Juli K. Dixon, Farshid Safi, and Erhan Selcuk Haciomeroglu
Develop a deep understanding of mathematics. With this user-friendly resource, high school teachers will explore strategies and techniques to effectively learn and teach significant mathematics concepts and provide all students with the precise, accurate information they need to achieve academic success.
BKF698

Making Sense of Mathematics for Teaching the Small Group
Juli K. Dixon, Lisa A. Brooks, Melissa R. Carli
Make sense of effective characteristics of K–5 small-group instruction in mathematics. Connect new understandings to classroom practice through the use of authentic classroom video of pulled small groups in action. Use the TQE (Tasks, Questions, Evidence) process to plan time effectively for small-group instruction.
BKF832

Visit SolutionTree.com or call 800.733.6786 to order.

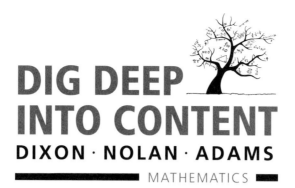

DIG DEEP INTO CONTENT
DIXON · NOLAN · ADAMS
MATHEMATICS

Bring Dixon Nolan Adams Mathematics experts to your school

Juli K. Dixon

Edward C. Nolan

Thomasenia Lott Adams

Janet B. Andreasen
Guy Barmoha
Lisa Brooks
Kristopher Childs
Craig Cullen
Brian Dean

Lakesia L. Dupree
Jennifer Eli
Erhan Selcuk Haciomeroglu
Tashana Howse
Stephanie Luke
Amanda Miller

Samantha Neff
George J. Roy
Farshid Safi
Jennifer Tobias
Taylar Wenzel

Our Services

1. Big-Picture Shifts in Content and Instruction
Introduce content-based strategies to transform teaching and advance learning.

2. Content Institutes
Build the capacity of teachers on important concepts and learning progressions for grades K–2, 3–5, 6–8, and 9–12 based upon the *Making Sense of Mathematics for Teaching* series.

3. Implementation Workshops
Support teachers to apply new strategies gained from Service 2 into instruction using the ten high-leverage team actions from the *Beyond the Common Core* series.

4. On-Site Support
Discover how to unpack learning progressions within and across teacher teams; focus teacher observations and evaluations on moving mathematics instruction forward; and support implementation of a focused, coherent, and rigorous curriculum.

Evidence of Effectiveness

Pasco County School District | Land O' Lakes, FL

Demographics
- 4,937 Teachers
- 68,904 Students
- 52% Free and reduced lunch

Discovery Education Benchmark Assessments

Grade	EOY 2014 % DE	EOY 2015 % DE
2	49%	66%
3	59%	72%
4	63%	70%
5	62%	75%

> "The River Ridge High School Geometry PLC went from ninth out of fourteen high schools in terms of Geometry EOC proficiency in 2013–2014 to first out of fourteen high schools in Pasco County, Florida, for the 2014–2015 school year."
>
> —Katia Clouse, Geometry PLC leader, River Ridge High School, New Port Richey, Florida

Contact your local representative
888.409.1682